Nicotine Rage

By Jason Aaron Baca

"Nicotine Rage," by Jason Aaron Baca. ISBN 978-1-60264-478-6 (softcover) 978-1-60264-479-3 (ebook).

Published 2009 by Virtualbookworm.com Publishing Inc., P.O. Box 9949, College Station, TX 77842, US.

Manufactured in the United States of America.

Special Thanks

A VERY SPECIAL THANKS to my loving and devoted wife Sheryl who has supported me every step of the way in my quest to become tobacco free. Also, thanks for being so patient with me.

I'd also like to thank my Mom who has dealt with anxiety & panic attacks for 9 long lasting years.

To Jeremy Kernion, M.S.W. who has guided our anxiety workshop every Monday and has made me realize that all things are possible even when faced with anxiety. To my sister Jessica and nephew Ethan who have continued to be my supporters and have kept the positive comments coming.

A special thanks to Father Gary Thomas who has prayed for me during times of depression and anxiety.

Richard Fanelli & Uncle Hugh who have both kept on me about depression and enlightening me about what I was going through. Without all of you as my support I'd still be using tobacco and be in worse shape then I was before.

To my friend Dave West, I never really knew what you were going through with depression until I got it myself. So now that I know, sorry for getting mad at you for all those years. I had no idea it was this hard to cope with.

Dr. Ryan Gorman, M.D., thanks for reassuring me that I wasn't going nuts all this time during my quit. If it wasn't for you explaining what was happening to me, I'd have checked into an insane asylum for sure.

A special thanks to my Grandpa for just being such a support. Grandpa, you are a great person to have on my side especially in times like these when I felt no one understood me. Sorry for keeping chewing tobacco a secret from you for all these years as I was trying to leave a good impression on you.

Preface

BECAUSE OF ALL THE ANGUISH AND STRESS nicotine and tobacco substance have caused on myself and wife, I've decided the best way to use all of this negative thinking is by writing it all in a book. My hopes are that both the quitter and surrounding loved ones can rest easy at night and not take things too serious in the first year of quitting chew.

I've spent nearly 1 year writing this book as I hope it will come in handy for those deciding to quit chewing tobacco. If you want to break a habit or addiction, this book is for you. If you want a guide from day to day feelings so that you know you are okay and not just having a nervous break down, then this book is for you.

If you want a book for your brother, boyfriend or husband that seems to always be leaving that can out where visitors can see it, this book is for you to secretly buy and leave on the table for him.

After going through all that I've gone through in quitting this dirty, raunchy habit, I hope you will reconsider quitting chew. Even if you aren't that thrilled about doing so, all I ask is that you read this on your off time.

My goal in writing this book is to possibly save a few human beings lives from this habit. It's considered one of the hardest habits to break and at moments I lost all respect for myself and thought nothing would ever be the same in my life. I was temporarily insane and thought, "I believe in God, but God doesn't believe in me."

My life today is pleasant, I feel good, the air smells fresh, I smile and it's easy for me to laugh at things again. I can workout and not feel a rapid irregular heartbeat now. I can visit with a large group of people and ask questions. My brain thinks now!

I attend holidays with my families all the time now and don't find ways to avoid get togethers. I take vacations to Maui with my wife and go to fancy restaurants. I go out of my way to make people feel good about themselves now which I never used to do before. I enjoy people once again.

If you so desire this and you are addicted to chew, you better keep reading because there may be some tools in here to get you back to feeling fine again.

Why Chewing Tobacco?

SO YOU'VE JUST COMPLETED a full afternoon's baseball practice for your college team and the coach made you run wind sprints at the end of the day to make sure he's got everyone in peak condition for the NCAA playoffs that you've finally reached. Playoffs start in a week but you are beginning to put things together in your head now on how you want it to play out.

You feel exhausted from all the focus and determination. Your mind is only on that big tournament and nothing more. As you make the short ride home back to the dorms you begin to unwind and relax but your mind continues to stay on the big games. The Coach explained that he wanted you to be eating, sleeping and dreaming about baseball and when you aren't doing one of those, you better keep your mind on baseball after that.

To better understand and visualize, you head on back to the dorm room to relax on your bed. As you do you lay down, you reach onto your coffee table and grab your can of Kodiak Wintergreen. You pack it with your hands so that you get a nice dense pinch. You put the chew in your mouth and begin your relaxation....You begin invisioning the score of your future playoff game and you being up to bat and going over each sense in your mind but as you do, you suddenly feel a bit anxious.

The anxiousness seems to get more powerful then just butterflies in your chest and it's actually causing you to feel a race in your heart and shortness in breath! You begin to think there is something serious wrong with you and your head begins to spiral. "Just what in the hell is going on here?" you say. It tapers off after a minute and you sit on your bed dazed and confused. You think to yourself, "Hey I've been chewing for 5 years and I don't see why it's causing me problems right now?"

Your roommate walks in the door and begins talking to you but you aren't giving him your undivided attention. Instead you are now caught up in your health and not knowing what's wrong. You leave the dorm and make it straight over to the nearest hospital where the Doctor quickly checks your blood pressure and heart. He does a complete checkup and says that you are fine. He tells you to just go home and rest and that you are in perfect health.

You think to yourself, "The doctor says you're cured, but you still feel the pain" It just doesn't make sense to you. Why in the world is this happening to you? And especially when you have some huge games coming up that they are going to need you for. You know all the scouts will be out there watching you and you need to get the butterflies out. You get back to the dorm and crash on your bed.

You realize that you are a senior in college and that this is it for you with college. And if you don't make a solid impression with at least one of the scouts, your baseball career could be finished. Pressure mounts for you.

The next morning you are woken 2 hours before your alarm clock sounds and you can't go back to sleep. You are nervous about something again and your breathing is shallow. You can feel your heart banging on your ribs but you are afraid to check your pulse in fear of it being way too high. You are screwed as you realize now that you are going to have to call the Coach and let him know you are sick and won't be able to participate in the playoffs.

You begin to break down and cry and feel like you could go crazy at any moment. You feel all alone because you've been waiting for this moment for a long time. The moment has arrived and these games decide if you'll be headed to Omaha, Nebraska, for the College World Series.

Now, because of your dip habit and addiction, this will remain only a dream for you.

You try to snap out of it and throw some cold water on your face and rinse off. But it's not working though.

You call your Coach to give him the news. The Coach has a hard time understanding but goes with it. He doesn't have time to hear your problems and will start someone else in your position. You can't believe you just called him and told him this!

Later that evening you receive word from one of your teammates that your team lost 8-3 but that the guy that played in your position went 4 for 4 and will be replacing you for Game 2. So even if you were feeling much better, you wouldn't be starting anyways. This also means that your professional baseball career is officially over since none of the scouts had a chance to take a look at you.

People look at you like you are crazy or that something serious is wrong with you which causes you even more anxiety.

Sad thing is after the anxiety is over, the depression usually follows. Remember, Anxiety is always anticipation of the future, while depression is always a thing of the past.

Yes, this is a typical anxiety attack. This is one that gets brought on from chewing tobacco. Get used to these sensations if you plan to become a regular chewer and once they start in, they take years to overcome. And once you decide to quit the chew, these symptoms get worse before they get better. I have a plan that can help you find your way to freedom in quitting chewing tobacco. My ideas are in this book. Let me help you get started in your journey to quit this nasty habit.

Facts about Chewing Tobacco

The Nicotine in Chewing Tobacco

THE AMOUNT OF NICOTINE you get from chew all depends on the type you chew, how much you put in your mouth, and how long you leave it in. So for example, you'll consume a lot more nicotine if you leave a dip in for 45 minutes as compared to how my Dad would do it for 10 minutes. Also, you can get a lot more nicotine from the average chew as compared to a cigarette. And you get way more of the chemicals (and that's the weakest brand, Skoal Bandits).

Tooth Problems

It causes changes in your mouth. I've heard from bodybuilders that chewed tobacco that they used to chew all the time while using "Performance Enhancing Drugs" and they claimed that their gums would become so tender that they could easily pull their teeth off. They believed that a combination of the 2 in the bloodstream caused this.

Also, after years or sometimes months, chew will cause your gums to recede back causing your teeth to be exposed more and more which will cause sensitivity.On a personal note, if you are reading this and you are a chewer still, switch to a “non-tartar control” toothpaste as the tartar control type will be a little too harsh on your gums or the area you put your chew in.

I learned this from speaking to my former dentist Dr.Stanley Wong (D.D.S.)

Mouth Cancer

A large percentage of chewers get white patches which are called Leukoplakia, better known as Pre-Cancer. This appeared a number of times along my gum line where if I was to rub my gums with my finger, the skin would actually peal away from my gums.

As you probably figured, people that chew are at a greater risk of getting mouth cancer, than those who don’t, which is a great reason to get off the stuff.

People often think that all chews are alike and contain the same substance. This is not true. There are different levels of chew that contain greater doses of nicotine and other "speed like" chemicals in them.

Strength of different brands from largest dose of nicotine to weakest—

- Kodiak
- Grizzley
- Copenhagen
- Skoal Straight
- Skoal brands
- Hawken
- Beechnut

The 7 can per week chewer takes in the same amount of nicotine as a person that smokes 4 packs per day! That means if you chew 7 cans a week then you are putting in the same amount of nicotine as the guy that smokes 28 packs per week. Sounds pretty unreal huh?

I don't know of too many people out there that are addicted to chew that only have 2 cans a week. On average, the majority of people who chew build a tolerance up to a can per day if not slightly more. This explains why it's so hard for people to quit this stuff! It's a lot harder then to quit then cigarettes because nobody can smoke 4 packs in a day!

The habit of snuff or chewing tobacco can be considered by many as a very nasty habit. The substance is held in the mouth. Usually on the side of the lower jaw and then spit out. Most chewers have a spit can or crystal geyser bottle that they carry around and spit into periodically. Souce - http://www.christianet.com/rehabtreatments/sideeffectsofchewingtobacco.htm

Some people may choose to engage in this habit because they believe it is not as bad as smoking cigarettes. However, this is not necessarily true. The substance is still absorbed into the bloodstream and the side effects of chewing tobacco can be very serious to health and can even lead to developing a serious illness such as cancer or even death… http://www.christianet.com/rehabtreatments/sideeffectsofchewingtobacco.htm

For those looking to quit, there is hope found in the Word of God:

"And Caleb stilled the people before Moses, and said, Let us go up at once, and possess it; for we are well able to overcome it" Numbers 13:30
Source - The Holy Bible

There are twenty-eight different carcinogens in snuff (and possibly even more not discovered yet) that are considered horrible for health or cancer-causing agents. As I was pointing out earlier, there is a larger dosage of nicotine in snuff tobacco then you'd find in a cigarette. There's actually a lot more in chew. They found this out a while ago and when they say "plug" they mean the weakest type of chew which would be Skoal Bandits. Some of the heavier, thicker based chews like Kodiak Wintergreen or Copenhagen have way larger doses of nicotine in them making them even that much harder to get off of.

In addition, there are chemicals such as formaldehyde (the chemical they use to preserve dead bodies) and arsenic (chemical in rat poison!) in snuff. Even without considering other chemicals the nicotine alone should be enough of a reason to stop chewing tobacco or the sensations should be powerful enough to scare you. Nicotine is a potent substance and has even been used as insecticide.

Nicotine crosses the blood-brain barrier and can be absorbed into the skin. Using the substance causes the body to increase the flow of adrenaline which causes your heart rate and blood pressure to go up as well as increasing insulin. The byproduct of nicotine can stay in the body for 48 hours after being used.It's possibly even longer for those that use chew more often.

Nicotine also increases dopamine levels in the brain which is why a person feels good when they use it. This is one of the main reasons that a person becomes addicted in chew. In addition, using snuff has been linked to developing Parkinson's and Alzheimer's disease. We'll get more into the damages chew causes later in the book. These risks are related to nicotine and all the chemicals that come with tobacco and how it affects the brain.

The best way to help someone with an addiction is to give him information that will educate them about the detriments to their body. There is a lot of information on the Internet that can be downloaded or requested through the mail by doing a search for side effects of chewing tobacco.

http://www.christianet.com/rehabtreatments/sideeffectsofchewingtobacco.htm

Here are a few more facts about chew that you may or may not know about yet.

It contains over 3,000 harmful chemicals including about 28 known carcinogens. Some of the chemicals in spit tobacco are:

- Polonium 210 (nuclear waste)
- Tobacco-specific N-nitrosamines or TSNAs (cancer-causing agents only found in tobacco)
- Formaldehyde (embalming fluid)
- Nicotine
- Cadmium (used in car batteries)
- Cyanide
- Arsenic (rat poison)
- Benzene
- Lead

Courtesy of:
http://www.tobaccofacts.org/tob_truth/spit.html

Arsenic is the one that gets me most riled up. I was watching forensic files the other day about a person who died of arsenic poisoning (or better known as Rat Poison) and how a person at work was apparently putting the chemical in his coffee everyday over a period of 6 months until he died. They were able to tell how long the arsenic was in his body by testing a piece of the victim's hair.

I guess my point here is, if arsenic is found in chewing tobacco then that means that it would have showed up in my body for 16+ years! That also means, say something was to have happened to me and I keeled over and died. Well in the autopsy of my body they'd be confused… They'd think I was poisoned! Or that someone purposely killed me! Someone would have gone to jail because of me!

This also explains why I'd have so many stomach aches all the time. During my 16 years of chew I'd always get that gut wrenching ache that would come and go and couldn't explain it.

I'm sure if you are sitting at home right now wondering if something happened to you and your wife was held responsible, you'd feel pretty bad too! So the next time you put a chew in, think about this and how you CAN actually hurt others in the chew process.

Like smoking tobacco, spit tobacco affects the cardiovascular system and may be associated with heart disease, stroke and high blood pressure. Long-term effects include leukoplakia, tooth abrasion, gum recession, gum and tooth disease, and loss of bone in the jaw.

Chew can cause cavity in your teeth and chronic bad breath.

It causes a lot of mental problems such as anxiety, depression and insomnia which I'll get into more, later.

You can even get pancreatic cancer and even liver cancer! Yes, I've discovered through taking a few specific liver blood tests (AST & ALT tests) that over the long term, chew can damage the liver badly and may even cause cancer in the liver some argue.

Chewing Tobacco is a Health Risk and can ruin your life!

Chew is dangerous to your health. People who chew also increase their risk of:

Can cause Mouth Cancer—cancer of the cheeks, gum, lips.

Chewing tobacco users are at higher risk of heart problems. And depending on how long they use, the numbers get even higher!

Can cause Throat Cancer—cancer of the voice box and cancer of the esophagus. You may find you'll constantly be humming before hand.

Here are a few problems that come with a chewing tobacco habit -

Chew can cause Heart disease—heart attack, stroke and high blood pressure. Over the period of time that I chewed I noticed my blood pressure would consistently get higher as the years would pass. I went from a guy that was athletic and fit and not at all worried about his health to a guy that would wake concerned he'd be facing a stroke at any moment.

Can cause dental diseases—yellow teeth, tooth decay, receding gum line, gum disease, bad breath and black tongue (as explained earlier)

Like smoking tobacco, spit tobacco affects the cardiovascular system and may be associated with heart disease, stroke and high blood pressure (Phillip Morris states that it "may be" associated with those problems but if you look at the numbers and all the people that have died from tobacco, I can pretty much assure you it's the main cause.) Long-term effects include leukoplakia, tooth abrasion, gum recession, gum and tooth disease, loss of bone in the jaw, cavity of teeth and chronic bad breath. It also causes a lot of mental problems such as anxiety, depression and insomnia.

Other health consequences of using spit tobacco include cancer of the mouth (including the lip, tongue, cheek and floor and roof of the mouth) and even liver! Yes, I've discovered through taking a few specific liver blood tests (AST & ALT tests) that over the long term, chew can damage the liver badly.

Chew can cause stomach problems—can even cause ulcers, stomach upset, increased bowel activity and stomach cancer. Be prepared for this stomach issue when you quit.

Chew can cause loss of taste and smell—causes loss of appetite which results in poor nutrition and poor health. Sometimes it got so bad, I couldn't even smell bacon cooking in the morning! Bacon!

Physical changes—fatigue, muscle weakness, dizziness and decreased motor skills. You feel clumsy more often and your reaction to certain things like a baseball being thrown at you, you don't react to the ball as fast as when you are fresh and not on nicotine substance. This is based on my own knowledge and not found out from a special study.

Chew can even stunt growth. Yes that's right! When I started chewing serious I was about 17 and in my prime for growing tall. But I ended up staying at 5'10". My Dad was 6'1" so I know I still had a ways to go. But they say a man can grow until age 22 but I stopped at age 17 because of this stuff. I know I would have loved to be a couple inches taller.

When you chew tobacco, you spread germs when spitting on the ground. I often thought that I wasn't causing any harm like smokers who blew smoke into the air but I was wrong. You can in fact pass on your cold to someone else just from spitting on the ground.

Leukoplakia – Now although for most it takes a few years. I read the Rick Bender story and he got cancer right away from Kodiak so a lot of it has to do with you as an individual.

Approximately 5% of diagnosed cases of leukoplakia develop into oral cancer. (We'll talk more specifically about Leukoplakia later in the book.)

Possible increased risk for heart disease, diabetes, and reproductive problems are being looked into but I can honestly tell you that from my years of doing this garbage that it does increase the risks of heart problems and diabetes . I remember my heart would always be under the control of the tobacco. I didn't know which way my heart would be headed.

Now a days (1 ½ years off the stuff) I have total control over my heart. When I go to the gym, it beats like it's suppose to and I don't feel like I'm going to faint.

After 6 months of being off chew,I had a lot of curious questions popping up in my head.

Chew can cause Leukoplakia (leathery white patches inside the mouth) are a result of direct contact with and continued irritation by tobacco in your lip. Just a few months of dipping or chewing is often enough to cause leukoplakia on the gum or cheek where the tobacco is held although for most it takes a few years. I read the Rick Bender story and he got cancer right away from Kodiak so a lot of it has to do with you as an individual.

Approximately 5 percent of diagnosed cases of leukoplakia develop into oral cancer. (We'll talk more specifically about Leukoplakia later in the book.)

Possible increased risk for heart disease, diabetes, and reproductive problems are being looked into but I can honestly tell you that from my years of doing this garbage that it does increase the risks of heart problems and diabetes. I remember my heart would always be under the control of the tobacco. I didn't know which way my heart would be headed.

I emailed Dr.Gorman who specializes in Internal Medicine and asked him if Chewing Tobacco can still play a roll in a person's vital signs even 6 months after a person quits cold turkey. His response was that yes nicotine is a powerful drug that attacks all the receptors in the body!

I also asked Dr. Thompson (D.D.S.) about how long nicotine can stay or play a roll on the mind and body. Dr. Thompson responded, "I have heard that nicotine can throw off a person's body chemistry for several years after quitting." A man who had used nicotine and went to him regularly had depression/anxiety issues for 3 years after usage.

Nicotine works at every cell in your body and when you use it over the years, you give it more and more power over you. Even though you enjoy it and it feels good being stored in your lip for 30 minutes, it is setting you up for a mental breakdown when you do decide to quit. If you think you can outsmart it too by moving the dip from one side of you lip to the other, that won't work either. It may give you a little buzz from not using in that part of your mouth before, but you'll still get sick, you'll still get mouth problems. If you stay in a casino long enough, its going to beat you.

When I began chewing

I REMEMBER MY FIRST CHEW, I was in Gresham, Oregon on a field trip with my friend's family. We were camping in the woods. I was going to be a freshman at Los Gatos High School. My friend Damon and I went hiking around the campsite. We noticed a car that had a can of Kodiak on the dash unopened.

We both looked at one another, both of us were curious about it since we both played and watched baseball regularly. We snatched the can and took off running into the woods. To us, we had found a new mystery, a new adventure. It was gold to us.

We came upon a grass field that was lit from the moon. We carefully opened the can and smelled the Wintergreen.. It had a strong, bold smell, the kind of smell that makes you quiver because of its power. "Here you do it first" I said to Damon. You could tell he wasn't prepared for the dare but went for it. He grabbed a nickel sized amount from the can with his thumb and index finger and placed it in between his lip and gums.

It was now my turn but after smelling how strong it was, I didn't know if I could do it...."C'mon, just do it," he said. So after a second I put in a good sized amount too.

After it was in my lip for the first couple minutes I thought, "This ain't so bad..." It just burned my lip basically. Then all of a sudden....Wham! It Hit! I got the most powerful buzz of my life. It felt as though I'd just had 5 shots of vodka. We both tried to stand up but we couldn't, we began getting dizzy and so we layed down on the grass. The dew from the grass felt good on my face. It was about the only thing that felt good in my entire anatomy as I'm recalling.

As I looked in the sky I can remember the stars were turning, I could feel nausea setting in and I began getting frightened. I thought this feeling was going to stay with me for the duration of the trip North to Washington. Damon was getting sick too. We both spit the chew out and swished our mouths at the nearest water faucet.

I remembered us both looking at each other saying, "What in the world was that all about?" Damon said, "I'm NEVER doing that again, it's just going to make me sick."

I believe he kept his word because all through out high school and even some years when I'd see him in college, I never once saw him with a chew in his mouth. Congrats to him for not touching it. I believe that that feeling that he had, had an everlasting effect on him. Now at the age of 36 I'm sure if he was offered some, he would remember that time he and I were in Oregon sick as dogs.

Me on the other hand, I forgot during my teen years. I'd say when I was around 17 years old I wanted to show my best friend Rich Fanelli that I knew something that would give him the biggest buzz in the world.It was even more of a buzz then alcohol. He didn't believe me and laughed at the idea of chew being this powerful.

So I said okay, lets meet at the Los Gatos Park and I'll show you! He agreed....We arrived at the park and both put a chew in our mouths. After 5 minutes he was in shock. He didn't know at all how powerful this stuff was. He kept saying Wooo as he would try walking around with it in his mouth.

Surprisingly though, he didn't get sick and only got the buzz but I got sick again following my buzz. I remembered feeling woozy and wanted to vomit. I felt off balance walking back to my car. Rich on the other hand loved every minute and had found his new thing.

A couple of weeks later at school I ran into Rich in the parking lot with a couple other friends. All 3 of them had big chews in their mouths and were laughing, and talking about sports. So I asked Rich where they'd gotten the chew from since they weren't even legal age to be buying the snuff yet. He said that the Liquor Store across the street sold it to them and that the company will take whatever they can get.

Apparently, I forgot how sick the chew made me a couple of weeks prior so I asked to join in the fun and have a chew as well. This time though, I didn't feel the sickness like I did before. Hmm, I thought this is actually pretty fun. Plus, I feel like such a big stud right now with this chew in and it's not even effecting me.

As time went on, I began chewing more and more often. I'd see a lot of guys in my high school chewing as well but I don't think they were as serious a chewers as Rich and I. We were up to 1 can a day by the end of High School. So in as little as 10 months we'd become full time, dedicated addicts.

It seemed that on every occasion that we had, a chew had to be part of the whole ordeal. Like for example, if we had gone to a party, we'd always make sure to have chew for it. Or if we went to the movies, we'd bring a can to have in the theatre. Both of us kept it a secret from our parents, especially because they were funding all this.

Nicotine had a way of making me feel calm in stressful situations, or so I thought. I know I handled myself fine before the habit but that was my excuse to use tobacco. This went along with the idea that I was creating this image of being a tough guy.

My Dad Chewed too...

EVERY YEAR MY DAD AND I would head down to our beach house in Cayucos, CA to get away. He was a regular Skoal Straight chewer. Although he didn't like seeing me use chew and was against it, he'd let me have one or 2 when I would go with him on the trips.

I think in his mind he felt that I was going to do it anyways so I may as well get sick from it when I was 18 rather then in my 20's. After a big BBQ I'd ask my Dad for a little tiny chew. No! He'd say. I kept on him. Common Dad just let me have a little pinch of it to try. Oh all right, but you are going to get sick Jay. He'd reply.

So I'd run out in the front yard of the house and open the can and delicately put a small chew in. I'd turn on my ghetto blaster and await my buzz. After a few minutes I'd begin feeling that dreamy feeling kick in and a dizzy head. Yea as you know, I'd get sick after that but I was so caught up in the buzz I'd originally get that it didn't matter anymore. He'd try to tell me when I would be sick and say, "See I told you this was bad for you but you didn't listen to me." I should have listened!

But then my argument was that how come he could chew a can a day and not feel a thing but I would always get these massive jolts? He said because he'd been doing it for so long and that this is what happens when you become addicted to chew. You don't feel it anymore.

We'd go fishing at the Cayucos pier and he'd pop open his can of Skoal Straight (Brown Can) and take a big wad out of place it directly on his tongue. Then with his tongue he'd push it down into his gum. He'd get it all over his mouth, tongue, teeth and lips. It was very disgusting. But I didn't think so back then. My Dad thought he was pretty cool with his Fresno State hat on, Blueblocker shades, and his big chew in.

I remembered just before bed at the beachhouse my Dad would have one last chew before turning out the lights. He'd sit and read the paper with one in. But when he was finished, he wouldn't brush his teeth or flush his mouth out. Over time, he'd lost some valuable teeth and the ones he was able to keep, were yellow.

My Dad ended up quitting chew 10 years or so later because he said it was getting too expensive and the higher it got, the easier the habit was to break for him. Later down the road though after he'd stopped, he reported having extreme anxiety, and bowel problems that would persist everyday. Besides those to things, he was constantly having to urinate and was having depression.

I noticed a big change though in his attitude over the first few years he'd quit. He was more of a calm guy and didn't get all angry at me all the time. It was a good change to see.

Now he's back to the same old Dad. He's been quit for about 12 years now I'd say. He still complains about stuff but now it's not as serious as before. Now he just complains that his steak isn't cooked enough at a restaurant or that the portions are too small or the waitress is a phony smiling all the time because she wants a big tip.

But I truly believe my Dad learned his lesson too about chewing tobacco without me having to ask him. It broke him down pretty bad in the long run and I remember many times he was scared he'd have a heart attack since that's the way my Grandfather (his Dad) passed on.

Now at the age of 61, my dad's body is beginning to break down. I'm sure even after all these years that the chew has something to do with how his body is working today.

Off To College

SO I WENT AWAY TO COLLEGE, which *really* gave me the freedom to chew whenever I wanted. Not only that, but I was now 18 years old and of legal age. I remember arriving for my first day of practice in college and I saw guys walking around with huge chews in their mouths. It almost made me feel like I needed to have a bigger chew in to prove that I was the real deal.

To be honest, I never liked chewing during a baseball game or practice because it always made my reaction time slower. It would screw me up when I'd come up to bat against a guy throwing in the low nineties and I'd be way behind on his fastball because of the chew. I remember this one guy who's name was "Bobby Brown" (not the musician) and he'd always be the BIG show off, making sure he had a big chew in, not talk to anyone but instead just smirk at them like he was too good to be there. Brown was a sure starter for our team from the start of the season. In batting practice he'd be the only guy using a wooden bat as if to say he's preparing for the big leagues.

Brown ended up signing with the Kansas City Royals as a non-drafted free agent and then played for several independent teams before hanging it up.

I can just imagine what kind of shape he's in now. He was an excellent outfielder, don't get me wrong, and I don't think the chew had any effect on his reactions in college but I do think as he got older, it had an effect on him. He was very quick and had a good jump on the ball back then but it does affect your motor skills. I'm sure it hit his eventually. It just had to.

This other guy on the team used to always ask me for rides back home after practice and would always try to mooch chews off me, his name was James Jones.

I will never forget his comments, he said, "once chew's got you, its got you for life "He'd chew Kodiak and tell others on the team that Copenhagen was the worst for you, then all the Skoal's, then Hawken, then Kodiak is suppose to be "the best for you" I believed him back then and thought his stories were from his extencive research or something intelligent.

I later found out that that statement was wrong and that Kodiak holds the most nicotine of any of the chews. In college baseball though everyone chewed, everyone! I hope Jones was able to stop. Everytime I'd see him, he'd have a big chew in. Hope he didn't burn a huge hole in his mouth by now.

I remember Coach Larry Lee would try being a good role model by not chewing and we even watched a video to scare us away from chew. The video would show guys that made it to the majors that chewed but had gotten mouth cancer so half of their cheek had to be cut off and a third of their tongue. This one guy you couldn't even recognize. He looked like a freak! I feel bad saying it about the guy but he does speeches around the USA about why you shouldn't chew. He would say things like "you don't want to end up looking like me."

Those videos tried to slow the team down from chewing and it may have worked for a few days, but soon the impression wore off in the players minds and it was back to buying a fresh can. For most guys on the team, it was what got them up at 6 a.m. for a long road game, and its what cooled them off after a tough loss and a long drive back.

The thing about chew, is it was for every emotion for a baseball player. When we won a big game we'd celebrate with a chew, when we'd lose a game, we'd be down in the dumps and pop a chew in thinking it would get you feeling positive again. I'd always try to visualize how my next game would be with a chew in and some music.

I just knew that after about one and a half hours, it was time for a chew. If I woke in the morning and didn't have any, I'd feel weird, kinda like a lot of people that didn't get a chance to have a cup of coffee or take a shower to start the day.

I remember all too well my daily regimen, I'd wake at 8:00 a.m. to my alarm clock, brush my teeth, drink a full glass of Crystal Geyser and then put a quarter of a can of Kodiak in my lip. From there I'd hope in the shower. I didn't need a cup to spit in since I was just spitting in the drain. I know it sounds gross, since I had two roommates that used this bathroom too. If you guys are reading this now, I'm truly sorry for being disgusting.

What I couldn't understand about myself during the college years is that the girls didn't like it at all. I think they were repulsed when they'd see me spit into a bottle or cup. I didn't care, I was more concerned with my image toward other baseball players and my addiction that I had on nicotine. To me, it was part of the jersey, and when I'd suit up, that would be the finishing touch.

So many times I'd go out to parties during college and I'd drink and feel fine but then would have a chew when I'd get home, and I'd end up feeling sick and hung over the next day. These hangovers would last an entire day too where I'd be sitting on the couch with a bucket.

College is a time when a lot of people find themselves, they find what they are going to be or be like. I found out that during my college years I went in with all the potential in the world of being a professional baseball player and came out nothing more then a guy that based his life around chewing tobacco. If it wasn't for the can, I'd have improved my game and been a lot more focused. Chew clouded my goals in my mind.

Not sure why I was the way I was. I'd listen to my elders and they'd say to get that crap out of your mouth or it's no good for you Jay but I wouldn't do what they said.

I remember how I went from a kind, soft spoken guy to a person with a bad attitude all the time. I found myself always complaining about something. "Oh the coach was rough on me today, why can't he be like that to the other guys?" I said. I never once thought of things how I would today. Today I'd think the coach was hard on me because he knew I had so much more potential then the other guys. Chew made me sensitive.

I remember playing a doubleheader in Bakersfield, California, and I was playing leftfield. This guy hit the ball way over my head and I was so focused on the ball, I completely forgot about the fence and I slammed into it. Dazed and confused I stumbled to my feet and located the ball. I tried to hit the cutoff man but missed causing the baserunner to get extra bases.

The coach stopped the game and brought everyone to him. He told me to get my head in the game. I was embarrassed as I've never been told this before by a coach. The rest of the game I played horrible because my mind was no longer on the game. It was on how upset I was. I wanted to go home and sulk with my chew.

Goes to show that chew doesn't make you calm at all though and it can really mess up.

The same thing happened with a teammate named Donnie Hill that was a 41st round draft pick by the Phillies. He chose not to sign and play a year in college for Larry Lee on my team. But I remember talking to him vividly one time. He lived right by me on campus and was cool when we'd talk off the field. He'd always have this huge chew in so he was "calm."

But then during practice, you weren't able to chew or have that stuff in your mouth, and if Coach Lee was to run into you with it, you'd be running laps for the rest of practice. But as I passed him in the dugout I waved and smiled as I do everyone and he just stared at me like a pitbull. Then he says, "Why do you always have that smirk on your face?" So I got worked up since I was being nice to him and I replied, "Why do you always have that pissed off look on your face?"

We both had a hard time keeping our cool for the rest of practice and think we both wanted to fight one another. But the point is, neither of us would have gotten mad or upset over little comments like that. Today if someone was to come up to me and say something rude like that, I'd probably laugh and I'm sure he'd be the same way, if he got off the junk that is. Donnie ended up never signing a contract from what I've looked up. He had a tremendous amount of pitching potential though and I'm sure it eats at him everyday. I'm sure somewhere along the line, chew had something to do with his career coming to an end so soon.

Somewhere along the line, he got upset with a coaches decision to not put him in or he walked off the field in disgust. Not because he was a hothead, but because the way Kodiak or any other chew makes you feel when you haven't had it for a few hours.

I recall one time at like 2 a.m. I hear a loud explosion outside which sounded like a brick being thrown through a window. Then I hear all this yelling. I look out my window and its Donnie being held back by a couple friends as he's trying to get at this one guy at our apartments. He kept pushing and shoving trying to get at this guy. I'm sure it wasn't anything big either. But the rage I saw in him that day was crazy.

All from chew, I guarantee it!

I remember getting mad at people that I had no business getting mad at. I wish I could turn back the clocks and say I'm sorry to them. I wish I could go back to that day in Bakersfield, CA and just take what Coach Lee said to me with a grain of salt. Maybe if I wasn't craving a chew so bad I would have! What a complete idiot I was. So I was still maturing as a man back then but that chew made me so immature.

I'd get in arguments with my roommates over the dumbest stuff like food or leaving clothes in the living room. I was not myself. I knew that I was addicted to nicotine but I wasn't aware that I was such a jerk!

What is nicotine addiction?

NICOTINE ADDICTION IN THE BEGINNING of my chewing years meant nothing to me really. It was an escape to me. Something I can do that I felt was more productive then stuffing my face with tons of food. To me, I looked healthy as a chewer because it constantly makes you go to the bathroom.

Later I would learn that that isn't such a good thing because if you constantly use the bathroom you are not getting the proper nutrients from the food and instead it just comes out as waste before your body and mind have a change to get their money's worth from it.

Nicotine addiction ruins many people's lives, marriages, family lives and just about any other thing you can imagine. You procrastinate constantly about doing vacations or trips. Before a fix you may be anticipating doing a nice long drive to somewhere peaceful. But five or 10 minutes into your nicotine habit, that trip doesn't seem so fun and instead seems like work to you. Nicotine would mellow me out way too far so that I would be so lazy that I wouldn't want to do anything for an hour or two and then when that two hours would go by I'd start thinking about wanting to go somewhere again but it would also be time for another chew so I'd choose the chewing tobacco.

The cycle would repeat itself and I'd end up bummed out cause I didn't do something fun all day. The nicotine also plays with people's brains in a strange way. One minute you could be feeling so happy about something, and the next, its no big deal.

When a person chews tobacco, the body responds immediately to the chemical nicotine in the tobacco. Based on what Dr. Thompson (D.D.S.) told me, Nicotine causes an increase in blood pressure. It also causes the arteries to narrow for a period of time and can "temporarily restrict arteries that flow to the heart." This, along with other nicotine effects, creates problems between the demand for oxygen by the cells and the amount of oxygen the blood it can provide.

So this kind of explains why I'd have a chew and about 15 minutes into it I could feel a tightiness in my chest or an ache.

Here is a list of some (not all) of the side effects that I personally felt from chewing tobacco for 16 years:

- fatigue
- laziness
- poor attitude
- lower motor skills
- hand-eye coordination was off

- developed negative attitude over time
- addictive traits
- heart palpitations
- dizzy
- insomnia
- shortness in breath
- tightness in chest
- indigestion
- constantly having to urinate
- bad breath
- grinding teeth when asleep
- antisocial
- light headed
- constantly having to "Hum"
- never getting a full nights rest
- procrastinate more often then not
- frustrated easily
- argumentative/difficulty thinking clearly
- low energy daily
- eventually lowers self-esteem
- feeling of jumpiness or nervousness
- emotional volatility, rapid emotional changes
- bad dreams

Now I'm pretty sure I left out about 1,000 other side effects but I know that I've talked to quite a few doctors and the majority of them agree that the above list is caused by the nicotine habit. I'll let you in on a big list later of all the anxiety symptoms that come with quitting chew.

I remember one Doctor saying to me that it affects every part of a person's body! So it is always working on breaking something down inside of you.

Anyways, here is a refresher again on why you should quit chew. I explained it earlier but feel the need to list it again. Skip over this section if you understand it already.

5 Big Reasons to quit Chewing Tobacco

1) Heart Disease

It causes Coronary Heart Disease which is a leading cause of death in the USA. The toxins in chew cause plaques to form in the arteries, which leads to hardening of the arteries. Chewing is hard on the heart as well and you are basically beating it up every time you put a chew in.

2) Stroke

According to the US Deptartment of Health & Human Services, stroke is the third leading cause of death in the USA, killing upwards of 150,000 people each year. For chewers the risk of a stroke is almost 4 times that of non-chewers.

3) Mouth Cancer

This stuff causes mouth cancer as we talked about earlier. The numbers are high that the use of smokeless tobacco can cause cancer in people. Most people that are chewing for multipule years uncontrollably find themselves with some kind of an issue with their mouth somewhere along the line.

4) Throat Problems

Every morning I'd wake up to mucus in my throat from over chewing the day before. For a good hour every morning you will find yourself clearing your throat. I was unable to find any reason for this from my research but I can say it sounds very rude when you are at a nice breakfast with your family!

5) Expensive!

Every year tobacco prices increase. When Rich and I began chewing tobacco in 1991 it was approximately $2.25 per can. Now in 2009, the average price per can will run anywhere from $7-$9 a can! That's almost a $10 a day habit! If you chew 6 cans a week, like I did, that would be $60 a week, or $240 a month, or $2880 a year! You also have to take into consideration your party nights that you may drink and tend to chew more on those nights adding to your bill. So I'd round it off to an even $3000 per year on chew alone would be accurate.

The biggest risk though is getting Leukoplakia! I don't really feel like putting photos of what people look like after contracting it but I can tell you, if you go on Google or Yahoo and type it in and click on search images, you will see for yourself. It is not pleasant to look at. Most of the people on there died from this as it turns to mouth cancer quite rapidly.

Crazy thing is it's hard to detect so a person that chews may have it and not even know. The way doctors get rid of it is by prescribing steroids to stimulate bone marrow into producting more neutrophils. It's a messed up situation and one that can be very scary. Another reason you don't want to mess with snuff.

I wouldn't want to be getting on steroids to get rid of a cancer because then you will end up with a whole new set of problems from the steroids. Steroids can also bring on heart problems and when you come off them depression can really play a big role in your world.

I look back at how much money I could have saved though and I could have been pretty well off! Ok so if we came to the fact that its $2880 per year and I did chew for 16 yrs that means I'd have had $46,080. Had this been in a 3 percent savings account, that's another $1382 (simple interest) Had it been compound interest than that figure would have been a lot higher! That right there is my down payment on a home that I could have had.

This is why I do an account that is totally dedicated to my chewing tobacco quit so that I can actually see those savings grow. This is also a reason to quit for those of you that like to accumulate money. If that's your motivation over health reasons, more power to you. Every 2 days I put $6 into an ingdirect.com account. I figure I was going to spend this on chew anyways, so why not put it into an account for fun.

Planning the Quit

I USED TO THINK THAT ANY TIME IS A GOOD TIME to quit but that was when I was still a chewer. I read through some websites that explain that it takes a "good couple of weeks" of withdrawals to get through nicotine withdrawal. I swear the people that wrote this are completely wrong and should be locked up for false information!

Quitting nicotine has got to be one of the hardest things I have ever attempted. And its not just having a few bad days here and there and then you are fine. No! You are looking at 6 months to a year if you were a heavy chewer.

For me, the deciding factor on quitting chew was when my wife and I were at Atlanta's airport getting ready to return to San Jose, California. I had just finished having a big chew in the terminal and we were boarding the plane. I remembered feeling hot, stuffy, and sick as soon as I stepped onto the plane. I got frantic thoughts that I wasn't going to be able to sit through this 4 ½ hour flight and that I'd go nuts. I didn't tell my wife I was having issues either. But hey, I had just had the chew so why was I getting like this?

It was at that moment that I said, NO MORE! That was it for me with chew. If it was unable to calm me down, then it was no longer something useful to me. This was once used as a tool to get me through the tough times but judging from this sensation, I was scared to ever use chew again in a potentially stressful situation.

Here is an article from Ian MG Smith who allowed me to use what he wrote about nicotine. This kind of gives a perspective of how challenging it really is:

Despite high IQ, good math, girls (secretly) liking me and being the top athlete in my age-group at school, I still felt inadequate, unpopular and a failure. Sound familiar, anyone? LOL

HOOK: Smoking and the attendant illusion of well-being became my coping mechanism. "For some people, smoking or chewing is a coping mechanism."

LINE: I remained an athlete and swimmer until my early 20s & 30s when I finally stopped all meaningful exercise and became a chain-smoking chair-jockey.

AND STINKER: By age 40 I was working up to 18 hours in my detached home/office, smoking up to 4 packs a day. (Please note that my good friend Ian was at the same level of nicotine as the average chewer that chews one can per day so his information is legitimate even though he was a smoker.) *I would only go back into the house in the wee hours of the morning for my single meal of the day and to sleep three or four hours. That was my life. How unfair on my dear wife and son.*

Eventually my brain became so affected that even the nicotine was struggling. Yep, it can wear off. Why do you think I needed 80 a day? Even that wasn't enough!

6 p.m. on September 25, 2007, I QUIT!

The next 3 months were filled with torment because I could not do my work. I became increasingly stressed of losing my job and the more I hammered myself, the worse it got.

January 2008 I hit the wall. BURNOUT! I was a shaking wreck, bedridden. The wife took three days off work and together we began to walk.We walked around the room at first, then the house and finally outside. Halfway around the block I stopped and felt the cold winter rain on my face. (Notice how Ian paced himself out and gradually became more confident again.)

That was the turning point!

Given the size of my addiction, I probably should have allowed at least 6 months quit for my brain to recover.

I lost my dear parents but (like Heather) stayed quit. Heather is so right, quit is quit-no-matter-what!

With exercise and a better diet, "Calms" from Karen (uk quitter) good sleep and positive thoughts, I got back on the forum and you guys were there for me. Hell, I ended up with the friends where we bragged about our new-found health and vitality and bathed in the afterglow of freshly released endorphines. Ha Ha Ha

HEY! I couldn't have done it without you all. September Enders, friends and everyone!

Finally, past few weeks (long time coming) the peace kind of crept up on me, deep and most profound, it brings strength and a new outlook on life. I am not a work-machine any more, I am a human being.

Karen (9 September) you weren't exaggerating. LOL

I am exercising and pushing weights again and the wife has taken up running. Hey, we even play badminton on Wednesdays.

Quitting has really boosted self-esteem and that feels good. Almost like my job again. I say it with a smile.

No ways can I cognite or perform like I did on nicotine (stimulant) but that was killing me. Instead of working myself to death I am learning to negotiate on human terms. This really is a whole new world for me!

Sure there are some aches and pains but I feel absolutely GREAT!

Heartfelt thanks to About.com, Terry, the Moderators, Pancake and ALL my quit-buddies.

Winston Churchill, 1942, "Now this is not the end. It is not even the beginning of the end. But it is, perhaps, the end of the beginning."

–Ian from About.com Smoking Cessation

Now this was an individual that was up to 4 packs a day and that was STILL less nicotine taken in then what I was doing in Kodiak chew. In the beginning though, you may think that you can get away without any withdrawal but think again! This stuff will eventually come up when you least expect it, when you have one of those days when you didn't sleep well the night before and then suddenly it hits you like a pile of bricks.

I dreaded certain periods of the day because I knew depression was bound to come up and did regularly around the usual time.

I never would get too high or too low. It was always this gloomy feeling. I noticed that whenever I traveled far from home, I'd get home sick and frantic. It was hell! No one ever talks about it either! I wanted the public to know how it changed my life.

I think the only thing that came good out of chewing is the fact that it brought me to believe in God. I used to never be one to talk about that kind of stuff either but now I make sure to thank him everyday for another day that I feel better from quitting chew. Chew had scared me so bad that I felt I had no other salvation other than Him.

"God is our refuge and strength, a very present help in trouble." Psalm 46:1

I relied on Him since I had tried everything imaginable to get better. I tried talking to Doctors but they all just look at you like they could care less about your problems. Or they'd tell you to take this pill or that pill to feel better. I tried talking to my family, but they didn't really understand what was going on inside my head.

I tried talking to my friend Dave who has been dealing with depression for the majority of his life. Dave was good to talk to for me and it felt good talking to a person that knew what you were feeling but then again, he didn't have the answers for me or he'd be doing great too right?

Friends are great to talk to but won't make you 100 percent better. Doctors don't care about you really and would rather not hear your problems and get you on medication so you shut up about it. Shrinks are okay but if you don't do the homework, you won't improve.

Group meetings of people that suffer from anxiety and panic attacks are quite helpful I've found. Depending on what kind of social worker you have lined up in the meetings, you get to hear other people's drama and the non-sense they have to go through as well. My group meetings had an instructor by the name of Jeremy Kernion, M.S.W. Jeremy was an excellent guy for the job because he never made you feel ashamed of what you did.

People would be in the class and complain about that they were having bad dreams, or were tired and his response was simple, "I don't care about the bad dreams, they aren't going to kill you," and, "No one ever died from being tired right."

He would tell us that he wants us to go out and search for ways to get anxiety. Like if people in class were always avoiding certain things, he would tell us, "Sometimes we all need to feel some discomfort in order to feel comfort." Before the class I never really thought about it that way. I always said no I'm not going to do this because I'll get an anxiety attack. But with this idea in mind I was about to stick myself out there.

You may think I'm getting away from the topic of nicotine and withdrawal but I'm not.If you are recovering from chew or smoke addiction, this anxiety stuff will involve you too so don't skip over these vital paragraphs. I paid for these classes, took notes, so that you can read them in my book. Learn it, Live it, Love it!

Anyways, back to the group classes, as I was saying, these classes are extremely helpful for those 1–5 months quit from nicotine because those damaged receptors in your brain are trying to re-organize themselves without the use of nicotine. I know I always used nicotine in any situation, happy or sad, this explains why when I quit I could never get too happy!

I remember the first class I took for anxiety there I was a nervous wreck. I didn't know what to expect. There were about 30 people in the class. When we all sat down, I found that I wasn't the only one that had a problem. There were others much worse then I so it took a lot of the heat off me.

One guy I recall kept moving his legs around nervously and his breathing was way off. Jeremy would go about the class room and have everyone say there name and favorite flavor of ice cream. I remembered not knowing a favorite flavor so I kind of got nervous as the circle came closer to me. I tightened up and heart was pumping hard. Then when it was my turn I just said, "Hi there, my name is Jason and my favorite ice cream is Chunky Monkey" (actually I think that was Jeremy's favorite).

But after my turn, I felt relieved, revitalized even because I dealt with the discomfort instead of fleeing. I knew that once I went to this class, it would be ideal for me since I had always shielded my social part of my life with nicotine or just avoided social situations all together which is something that nicotine is good at making people do … avoid!

Jeremy would go around the room each week and ask people what they did for the week. Each one varied from what they did. Some had it so rough that they couldn't go to a grocery store and say Hi to the check out clerk. And here I thought I was going to turn heads with my problems.

The instructor found ways with each person to make him/her feel good about their current situation. Like for the guy that had a hard time of it in the grocery store, Jeremy explained that he was just proud that the gentleman went into the store and that that was brave of him. He also noted how strong it was for him to come to the meeting that night. I agree that meeting can make people uncomfortable because you will be talking in front of a group of people and not just one on one.

So yes, I would have to say that taking an anxiety class or 2 would be a good thing and you can only get positives from going. Even if you don't really have problems in general, it's a fun class.

My friend who has Bipolar disorder had been going to group meetings for over 10 years. I remembered how I'd ask if he wanted to meet me to play a game of basketball and he'd say, "No sorry, I've got group tonight." One day I decided I'd go with him to see what he kept talking about. After going to his class I realized why he went so much. And even though I didn't have Bipolar disorder I could understand a little more about his problems.

He was in the same boat as I was with the anxiety. He wasn't nearly as bad as the others in the class even though he may have thought that when he first started going. The group was designed similarly to the one I attended only Dave's class dealt with a disorder that will stay in a person's life forever whereas mine had the potential to disappear if I did my homework.

This was the only time in a 10 year period that I had attended a group meeting. I had not experienced depression or anxiety before quitting chewing tobacco. As soon as I quit, I realized why Dave had been attending classes.It's because of how good it makes a person feel, hearing others mope about there hard times. I don't understand why it makes a person feel good, it just "does."

So if you are like me after you are quit, and are feeling melancholy, group classes are my advice.

Days before the quit

Before you quit, announce to the world that you are quitting tobacco. Not just to one or 2 people but make phone calls, write letters and stick to it....Say on 7/08/2008 I will quit tobacco after 16 years, for good. I am going to do it cold turkey (or whatever method works for you) and stick to your word. Don't let them down. Make sure this quit is coming from you and not a nagging spouse.

My mom used to always hassle my stepdad about quitting smoking. He had smoked 2 packs a day since he was 16 and was now 56. That's forty years of his life that he's done them. BUT, it had to be up to him if he was going to quit or not. He had tried several times after my Mom told him to and they never lasted more then a month or two.

This last time he quit though, you could tell it was coming from him wanting to. I think he had a scare of some sort that he never told me about that caused him to do so.

He seems a lot more in tune now that he's done smoking. He doesn't get as angry that easily either. Now when he's done eating, he falls asleep instead of smoking which isn't good for you either but would rather he do that for enjoyment then smoke.

A lot of people on the smoking cessation forums announce they are quitting a couple days before and then a few of the moderators will follow up with these people to see if they stuck to their word. Often times the moderator will reply to the post and the person won't respond. They usually are a good indication that they changed their mind.

Celebrity Rehab, VH1 Part II

So I was hooked on a reality show titled *Celebrity Rehab*.

The show, *Celebrity Rehab 2*, starring resident Dr. Drew Pinsky, aired on VH1 in October 2008.

Now, when I first tuned into the show I was very interested in learning how they would be in a treatment center. It felt good actually to see others going through a hard time with whatever their device was they were hooked on. I felt I could relate to them. But the part that turned me off from the show was the fact that they allowed the people to smoke cigarettes. It got to the point that whenever you'd see them talk outside, they would be holding a cigarette. So it's like they are defeating the purpose of being there!

They get off pain pills or alcohol only to find themselves addicted to nicotine more then ever now. Does this make sense to you? I asked this very question to people in a quit smoking online group and some said that "at least they were quitting a big drug."

They were making light of nicotine from what I could see. To me that's like telling an alcoholic that it's okay to have some wine but just don't drink Vodka. You are still getting the stimulant somehow. Well I nearly got kicked out of the group for voicing my opinion. Apparently one person had lost a loved one to meth and said how she went to a treatment center and they told her she could smoke instead. She ended up dying somehow from the stuff. It was a sad story and one that caused the whole forum to turn against me.

Anyways, about the show, Dr. Drew does excellent work in getting the addicts to speak about their habit. He monitors them, asks them plenty of questions to find out what got them started in the drugs and then he lets them talk amongst themselves.

The other part that I am confused about is that after three weeks of rehab, they go through a "graduation" from the drug. It seems to me that with only 3 weeks into the quit, people are just barely getting started living in their new self. I know that was the case for me with chewing tobacco. I know of people that have been hooked on meth and they said it took them about 1 year before they could function somewhat normal again.

It just seemed like a big joke to me. The celebrities come in there all messed up and they show footage of them as they were addicts. Then they show them graduating, dressed up, and hair combed.

If they really wanted to do a rehab show, they should do taping on and off for half a year. Edit the boring days and use the ones that show them crying, losing control, vomiting, having panic attacks, being stuck on the toilet, losing balance when walking, contemplating suicide, or not being able to sleep at night. Do that show, and you'll have a bigger audience VH1.

Resources for Quitters

THERE ARE NOT TOO MANY PLACES a person can find resources for those that are interested in quitting chewing tobacco. It is such a special audience that I think most books would sit on a Barnes & Noble shelf forever. And yet, I needed a resource book for chew. I felt so alone in the beginning and ended up having to read quit smoking books instead.

I even went up to an available bookseller at Barnes & Noble in Campbell, California, and asked them where I can find a book in their store on chewing tobacco. She looked at me like I was speaking a foreign language. "For what exactly?" she said. I said for quitting chewing tobacco? Do you have books where if someone wanted to quit, they could read this book? "Oh no I'm sorry "she said. "We don't carry those kinds of books at Barnes & Noble so you may want to try a library"

Of course the library was a dead end too and I don't think the librarians ever heard of the term "chewing tobacco" in their lives. And it's hard trying to explain it to an old lady in her seventies.

There aren't many quit smoking books either if you go to a local library or bookstore. Almost everything in the "addiction/recovery" section has to do with pain medication, heroin, cocaine, alcohol and other illegal drugs.

I never understood why either since nicotine is just as addicting and hard to quit as cocaine and heroin. Look it up online and I'm sure you'll read all about it.

As far as my search goes, I recommend going to about.com and typing in "nicotine addiction." There you will find resources to get you started on quitting. I also recommend going to killthecan.org which is for chewers. A great site that explains a lot in detail but isn't correct on a few topics like how long it takes to withdrawal from it. They claim in a couple weeks you'll be fine. That is only if you were an amateur chewer that did it for a year or so and had maybe 1 chew a day.

Some of these guys in there aren't serious and you can tell from the way they talk tough and say that chew has never affected them. Ignore them, they'll have their turn, trust me!

Also an Excellent resource is a site called "quit smokeless" quitsmokeless.org. Let me share with you the opening page to Matt's website as it is one of the few sites that specifically targets chew addicts.

Hi, I'm Matt, like you I have battled with chewing tobacco and about every other kind of smokeless tobacco that exists. It's now been almost 10 years since I have had any tobacco. I chewed for 8 years (I'm 33) and probably tried quitting for the last 5 of those years. I tried the patch, gum, etc... To tell you the truth I don't know how I quit! It wasn't any one thing, but the combination of motivation with circumstances made it happen.

"So how does this help me?" you may be saying. Basically, I started this page because, one, There isn't jack on the Web (or anywhere really) supporting those trying to quit this kind of tobacco addiction, and, two, I think about how much friggin' money and time I wasted on the crap and I wish someone who had been there would have told me what was up! I guess I wouldn't have listened anyway. How many times have you made lame excuses to leave your girlfriend just to slap in a wedge and sit around filling up a pop can with spit? No less than a hundred for me. But, that's just one of the problems I had. I don't think I'll get oral cancer, but they (the medical community) say it can show up even years after quitting.

Oh yeah, I haven't really stated my point. If you want to quit, eventually you will, provided you never quit trying to quit. That's a mouthful. I don't know the answer to what works best for quitting, but I think having people care about your well-being is an excellent start. I was fortunate that I had guy friends who wanted to see me quit as well as my girlfriend, now wife. Also, replacing that craving energy with physical activity really helped me through the withdrawal. I started running and weight lifting. Oh, and I ate sunflower seeds like a freak. I'll take a little more salt over cancer.

Why would we want to quit using smokeless tobacco? There are obvious reasons, and not so obvious ones. Let's review the obvious ones first:

- It can cause cancer.
- It has you under its thumb. Why else would you drive to the Seven-Eleven at 3 a.m. for a tin?
- It costs too much money.
- Other people don't like it as much as you do. (Some people don't think that thick coffee spit is cool, like I do?)

Okay we all know that it is not healthy, and that's ultimately why most of us even think about quitting. But what about the other stuff that nags at us? I'm talking about the friends, family, wife, girlfriend, partner, whoever, who keep hassling you to quit. Do we care? I guess it is hard not to. After all, the quitting thought crosses your mind a lot anyway, so their concern isn't a surprise. Unfortunately, that concern can make you retreat to the tin or pouch to contemplate the whole situation over a big fatty in your cheek or lip.

Guilt sucks.Feeling guilty is one of the less obvious reasons to quit, but it certainly is one of the strongest. On the other hand, those guilty feelings become a good thing. They keep reminding you of what you are working toward.

Basically, you're here because you want to quit, and that is good.

You will quit. Not necessarily today, tomorrow, or even this year. The fact that quitting is on your mind at all indicates you have some common sense. Guilt will hassle you every so often, you'll try to quit again, and you'll succeed or you will fail. If you're successful, you are done. If you fail, you'll try again. Someday, you will quit.

The bottom line, as I said before, is keep quitting until it works. I've seen some Web sites from guys who are really into their Grizzly or Copenhagen, or whatever and they proudly display pyramids of tins that they've dipped. That's fine, I did the same thing. But now, it just seems odd to be all fired up about a product that costs you over $1000 (even more now since prices of a can of chew have gone up since this report) a year AND is hell to quit. Anyway, I have rambled on enough for now. Thanks for reading this and I hope you keep up the effort.

Courtesy of Quitsmokeless.org

Getting out of Dipville

By Goodbyecope © 2003

DIPVILLE IS A DISGUSTING PLACE full of disease, slow but certain death, lies, deceit, and filth. The residents of Dipville appear to be normal to each other, but rather strange to outsiders. You see, the residents of Dipville all have this habit of frequently packing their lower lips full of this dirt-like substance. This substance goes by many names: dip, snuff, and chew to name a few. It comes in a convenient round can about the size of a hockey puck and is to the resident of Dipville a necessity, a life-giving substance, something to plan one's entire day/life around. In a sense these cans are very much the Dipvillian's best friend and yet at the same time their worst friend. To anybody living outside of Dipville, this practice is disgusting and foul and the outsiders think the residents of Dipville are quite pathetic and sad and gross.

People have a strange awareness of always knowing where their little can is at all times. They always have an inventory in their heads about how much dirt they have left in the can and the approximate time that it will run out. This way, they can time their stop to purchase a brand-new can of this life-sustaining dirt just in time for the old can to run out. There is a ritual to breaking the seal on a new can, packing the contents and stuffing the lip. There are variations on the procedures, but it's still a constant among all residents that it is a daily ritual performed several times during the day.

There are a few problems with the tobacco that must be explained. Beyond the fact that it creates brown ooze in the Dipvillians' mouths that must be either spat or swallowed, there are much more serious aspects. This dirt causes the mind of the Dipvillian to become extremely dependent on it. Not remembering to insert the cope into his mouth will make a Dipvillian dizzy, mean, cranky, and just plain sick. This is one of the ways that residents are encouraged to stay in Dipville. The dirt creates painful sores in the mouth. These sores often turn into lesions and eventually (with enough dirt) will turn into cancer. Mouth and throat cancer is almost always a death sentence. Needless to say, there are many Dipvillians who will get this cancer if they don't figure out a way to get out of Dipville.

Some of the residents of Dipville really live just beyond the city. They themselves are as much residents as the others, but the difference is that their families don't know that they frequently sneak into town to be with their can and stuff their lips with dirt. They lie through their dirt stained teeth just to have some time in Dipville. Some of the residents have been living in Dipville since they were 10 or 11 years old. It's no surprise to find residents of Dipville that have been there for 30+ years.

It is incredibly expensive to live in Dipville. Just buying a fresh can of dirt each day can cost the Dipvillian anywhere from $120-$150 per month. The cost is increased by needing to continually purchase new containers to spit the brown ooze into. Frequently Dipvillians buy a soda each day just for the container to deposit the ooze. This increases the cost of living another $30-$40 per month. Increased use of gasoline to make strange trips to the store, bank, etc. (just to have a mouth of dirt) can add another $20–30 per month. So for the average Dipvillian, their lifestyle costs them from $170-$220 every month. Couple this with the cancer risk, and there's a real sacrifice being made to remain in Dipville.

Most of the residents of Dipville have tried to leave many times, only to find the move too difficult. They make-up excuses about how difficult and trying life is outside of Dipville, or that they just want to take a short ride back through Dipville and have just one more can (which almost always turns into 10, 50, 100 cans...) Or they take a little trip into Smokeville and the little sticks of dirt there remind them of the little cans of dirt in Dipville and before they realize it, they are right back in Dipville enslaved to their old ways.

Most of the residents of Dipville hate living there. They are tired of the smell, tired of the slimy brown ooze coming out of their mouths, tired of talking to people with a lump in their mouth, tired of the lies they tell to their loved ones just so they can remain living in Dipville. They are tired of their mouths hurting, tired of wasting their money on cans of arsenic, tired of being a slave to the little cans, tired of hiding. They are just plain tired of living, breathing and slowly dying in Dipville and want to move out for good. Countless numbers of the residents have tried to move away, but have come right back to their old homes in Dipville.

The best/worst thing about living in Dipville is the bridge. This is not an ordinary bridge, but rather it is the bridge to Freedomville, a wonderful place that is actually visible from Dipville. Freedomville is so beautiful. Nobody uses dirt in Freedomville, nobody has brown ooze coming out of their mouths in Freedomville, and nobody is chained to a can in Freedomville. They don't have to lie about dirt in Freedomville, they don't waste their money on dirt in Freedomville, people don't make up errands to run so they can stuff their lips with dirt in Freedomville.

Some of the residents of Freedomville have lived there all their lives. Some used to live in Dipville and have relocated permanently to Freedomville.

And there's the sad crowd of those that only live in Freedomville for a few weeks, days, hours, minutes at a time before crossing the bridge back into Dipville. They cross the bridge into Freedomville, try to assimilate, and try to stay out of Dipville, but they're missing something. They still see the bridge to Dipville as being open to them. They remember that the can of dirt they used to love is right there in Dipville just waiting for them. They remember all the activities they used to do with their can at their side. What they don't do is see themselves as full-fledged residents of Freedomville. They don't look at the bridge and remember that "The road to Dipville is off-limits today."

Fellow quitters, we're in Freedomville right now. The bridge to Dipville is there. But we must believe that today we are truly residents of Freedomville. We must believe that the bridge is off limits to us today. We must believe that it is fully within our control to not cross the bridge to Dipville today. We must remember how disgusting Dipville really is. We must believe that we will never drive through Dipville again for the rest of our lives. We must remember that the can we loved was going to kill us. We must remember that in Dipville we become dishonest and ugly, un-trustworthy and scheming, we become slaves to the dirt of Dipville.

The stresses of daily living will come our way every day. Let's face those stresses in

Freedomville, not Dipville. No crossing the bridge today! Dipville is off limits!

I want to thank Mr.Young and quitsmokeless.org for allowing me to use this very true story from their site.

These are a couple of my favorite stories that are on quitsmokeless.org. There are many others on Matt's site but you will just have to go there and see for yourself. He has a community as well where you can talk to others about your experiences. Not only that, he has a cancer photo gallery which is not for the faint at heart. I was never one to like looking at blood and guts, but this you should look at if you are considering staying on chew… Some of these guys chewed so much that their bottom lip is as large as a golf ball. If this happens to you some day, then you can't say I didn't warn you ahead of time. He also has a "quit tracker" that shows you how long you've been quit, and how much money you've saved by now chewing. So for example, the guy that quit 191 days ago has saved himself $573.00 already! It's a fun site and you can get lost in it if you are a chewer contemplating quitting.

You get the point that this is a great site. I hope that by the time you quit, that this site will still be out there.

Courtesy of Quitsmokeless.org

How important is sleep?

In Dr. James Maas's book "Power Sleep" he talks about what can happen to a person if they don't get enough of it. I was in complete shock when I realized how much stuff can go wrong from depression to low concentration levels. He does extensive research on the topic and explains how tobacco usage can play a huge roll in insomnia.

Insomnia can get brought on by chewing tobacco and smoking because it's a stimulant. People think it calms them down but it actually speeds them up. I used to chew before bed and it would end up taking me 3 hours before I could fall asleep at night. The result — I'd become sleep deprived. It's not fun when you are trying to go to work and you're not fresh in the morning. That's no good. You want a full nights rest.

When I quit the chew I noticed that that was a severe withdrawal symptom that took a lot of patience to work through. I'd want to sleep during the day but would force myself to stay awake and then at bedtime I'd be wide awake! The cycle repeated itself from month's three to six in my quit.

On the about.com smoking cessation forum a lot of people complain early on in their quit that they have sleep problems. One person noted–

I didn't sleep for the first week or two. I could fall asleep but couldn't stay asleep. I put it down to my nerves being raw (without the nicotine to deaden them) and so any little thing would wake me from sleep. Who knows, I just know it's really hard to stay in a good mood when you're so tired.

That last sentence is a very good point because it IS really hard to stay positive when you are tired. You think mainly negative thoughts and they don't seem to leave your head. It is so frustrating. This is when a lot of people turn to meds for help. There is nothing worse then feeling mentally tired & physically restless. Your brain is craving sleep, but your body says NO. It is sometimes a very long process. You can go to bed at night feeling like you are so tired that you will have no problem sleeping 8 hours. Then you find yourself waking up halfway through the night wide awake with an upset stomach.

Medication for withdrawals

Ok I must first say that I'm not a Doctor. What worked for me might not work for you. This is all my opinion. My theories here are written to be challenged.

I have attempted many crazy pills during the quit. I attempted Zoloft, Xanax, Trazadone, Celexa, and Wellbutrin. I found that when I started each one, they all made me worse off. My mind was very sensitive to foreign substance entering my body. I didn't like feeling like I was in slow motion constantly.

I knew when I quit chew that medication would be my LAST resort and would only take pills if I was desperate. There were five occasions where I felt hopeless in quitting chew and felt there was no other way about it but to take pills.

Every time though, I'd end up taking the pill and being even more freaked out as a result. Therefore, I made a pact with myself and that was to "ride it out" naturally. I was going to let my brain correct itself on its own without the help of an SSRI or beta blocker.

How could I keep track of my progress if I was taking meds?, I asked. I felt that if I started feeling better, it would have been because of medication and not because I was healing naturally. I read somewhere though that meds aren't a gateway to a new life though and you still have to work at improving yourself and building your confidence back. I have friends that have been on meds for a few years now that tell me this too.

I've seen guys in my anxiety classes that come in there and were extremely nervous and you can tell they were having anxiety or panic attacks. Then after a couple weeks of them doing a med they'd calm down. So yes, it works for a lot of people or meds would never be available if people just jumped off bridges when they did them.

Let me share another story from the quit smokeless site.

What Price to Save Ourselves?
by Spongebob © 11/20/2002

For over 2 decades, my best quit efforts lasted maybe 10 or so days. So basically by asking myself the right question it changed my attitude and made it possible to quit. This quit is not easy, but it is finally in MY CONTROL and (I firmly believe) FINAL.

Previously, I always asked "how can I find the strength to break this addiction? In particular, how can I get through the crushing brain fog that always leads to my my returning to chew? I can't stay quit or start quitting right now because I get too brain-stupid to get any work done." THAT QUESTION ALWAYS LED ME TO GO BACK TO CHEW because (a) It gave me the choice to simply start over and (b) It said I had other priorities that I would allow to interfere with quitting the can.

This time, I asked myself a different question. "IS THERE ANYTHING I WILL NOT DO IN ORDER TO QUIT? IS ANY COST TOO HIGH?" Since nobody was asking me to give up my loved ones and friends, I decided the answer was "NO." I therefore decided that I would do whatever it took to quit this stuff. If I must, I will use up all my vacation time to get away from the office until the anxiety fades or depression passes. If I have no vacation time left, I'll call in sick or use the rest of my floating holidays (and I consider addiction withdrawal to be honestly sick). If I run out of vacation/sick time, I will ask for unpaid leave (better known as a FMLA) until my head clears up and while I practice handling fewer stresses without opening a tin. If I can't get unpaid leave, I'll let that job go (and go find a new job AFTER I SAVED MY LIFE). If I can't afford being on unpaid leave or unemployment, I will swallow my pride and ask for help from family, and I will sell my car/house/stereo. The point is, none of these items you can't replace. The only thing you can not replace is your life!

WOW, once I decided that NO COST WAS TOO HIGH TO SAVE MY LIFE, and that I would GLADLY INCUR THOSE COSTS, my whole thought process changed in a matter of minutes. No longer were there any impediments to quitting. Once that was my attitude, quitting was better than I had experienced in prior efforts. I did have to cut back on my office time either (and incur some temporary pay reduction) but nothing too crazy. And in the long run, who gives a damn right? I've got my life back and that's all that should matter.

See, the real barrier wasn't quitting tobacco — the real barrier had been what I had not thought about doing, or had not been "ready" to do, in order to make quitting the absolute #1 priority.

Another example: does quitting make being around the house unbearable? Negotiate leaving for 2 months!!! (of course as you'll later learn, it takes a heck of a lot longer then 2 months to get over this chew habit, this guy had it lucky compared to most.)

"Honey, I need these 2 months in order to give you the rest of my life. This isn't a vacation, this is the old 'stick with me in sickness and in health thing.It is unfair to leave you, but I will make it up to you, and you will like the new me 10 times better, and I won't go and die on you.

I came to this "At What Price" attitude after my wife died. She had been given a terminal diagnosis from hell with no hope whatsoever (Her sickness was known as Lou Gehrig's disease). We had wished there was something we could do, but there was not. And she had done nothing to deserve it (she didn't smoke, barely drank, exercised regularly, healthy).

Now here I was, 14 months after she passed away, giving myself my own death sentence. But this was a sentence I had the power to stop. My wife had been denied any such power. So, every time I CHOSE to fill my mouth with chew, I insulted the memory of my wife. My wife and I would have paid ANY PRICE to save her: sacrificing job, etc.

Once I asked "What Price?" to save myself, the answers became rather obvious and easy. This quitting chew thing is not easy, but now it is only a question of time. The fog and depression still hangs around some, but now I just ride it out rather than fight it or let it scare me back to the can.

It's been 15 months and I've come to terms with my loss of my loving wife. But I wanted to share this story to prompt you to ask yourselves, "Is Any Price Too High?" Are you putting fake barriers (like the job, or conserving vacation days and sick time, or keeping things from your wife) in the way of accomplishing THE MOST IMPORTANT GOAL in your life right now?

Would you quit your job, sell your house and move to a desolate place where you have no friends, all in order to save the life of your child, wife, or father? Of course you would. Now, do whatever you have to do, at whatever cost, to save your own life.

Spongebob is right; accomplishing quitting tobacco IS the most important goal in a person's life and should be for as long as it takes to feel the withdrawals. I liked how he put that. If you were to put a person from your family on the line, how would things change? You'd be surprised how strong you can become in an instant! He went through quite a bit losing his wife to Lou Gehrig's but like he said and I think the attitude he possessed in quitting chew, is the same attitude you must have going into your quit. There are no options to putting the stuff in your mouth, none!

This next article comes most importantly to the wives of husbands that chew tobacco. I'd have wanted my wife to read this too as it is very good indeed.

Courtesy of quitsmokeless.org

An Open Letter to Wives of Smokeless Tobacco Users by Rick in Tampa © 2002

First thing I want to say is "Please don't get discouraged"

Regardless of who says what to you, there is no single way to force your boyfriend or husband to quit using Kodiak or Chewing Tobacco. No ultimatum, no New Years Resolution, not a child being born, not the threat of you leaving him will honestly convince him it is time to quit tobacco. I apologize for being blunt but its God's honest truth. Every person has that little switch inside him that tells them when its time to stop. Your chore is to find out how to flip that mind switch of his. I can tell you that, you have found the single best resource I have ever found to help your boyfriend/husband quit smokeless tobacco

I started chewing tobacco when I was 13 years old. When I quit 100 days ago, I was 'using' a can of Kodiak Wintergreen a day @ $4.55/can. (please note that when he bought chew it was in 2002 and a lot cheaper, today in 2009, it ranges from $7–10 per can!) *I would stay awake until 1–2 a.m. so I could keep dipping. I would brush my teeth in the morning only and as soon as I was done, I would throw in a dip before I hopped into* *the shower. I hated brushing my teeth because it hurt. The Tartar Control Toothpaste was too harsh on my gums.*

There are a lot of things about the smokeless tobacco addiction that an addict will never admit to and will never share with anyone else, its part of the "being addicted to a substance."

In my life I never watched a movie, mowed my lawn, washed a automobile, drove a car, sat at a computer, went to the bathroom, sat outside, played baseball, football or a round of golf without a dip in my mouth. Sorry if that's too much information but it's the truth. Sometimes the truth hurts but there are a lot of people that are the same way that take on chew.

The reason I didn't want to quit or was scared to try is that Kodiak Wintergreen defined who I was. It was part of my person. I felt incomplete without it. I had no idea how in the world I would do 99% of my daily activities without it.I am married and have 2 kids and nothing that my wife did not even the birth of my boys inspired me to quit.

I wish I could explain what it was, but after 20 years of chewing, I just decided it was time to give it up. I had moments like that before, where I felt this is the last chew I will ever have but it lasted a mere hours. Then I would go buy another can or have another chew and start over. This time was different because I found the Quit Smokeless web site. I posted out there on Monday February 25, 2002, for the first time and I haven't had a dip since. I am convinced I will never have another one again, for the rest of my life, period.

But for me, I had to get serious. I had to change a lot. I stopped eating unhealthy food, stopped drinking coffee and Coke ('pop' as we called in Pittsburgh) and joined a gym that same day. I have not missed a workout since then and have maintained my weight but got rid of most of my gut. I don't pretend to have a special touch or that people follow my 'quit plan' because each person quits in different ways for different reasons. This happened to work for me specifically.

The benefits are nothing short of incredible. My teeth feel strong and healthy again. I sleep less number of hours, but wake up feeling more refreshed than I ever did when I dipped. My tongue isn't brown from dipping. My fingers aren't stained. I bet we go through 1/2 of the toilet paper now that I don't stuff that in a cup every time I clean out an old spit cup. My car does not need to be cleaned of all the tobacco that would fall out of my can every time I took a dip while trying to drive at the same time. Food tastes like food and I don't treat meals like its something I do to survive between dips. My gums don't bleed anymore when I brush my teeth and I can actually use Tartar Control or Whitening toothpastes again. I spend more time with my kids and have less anxiety about 'when is my next chew going to be. I feel less irritated and more jubilee. And I kiss my wife more often. I just can't explain how many ways I feel better.

I didn't write this for vanity purposes or to show off, I tell you them in hopes that your husband will read this and see that quitting is possible. If someone would have told me this a year ago, I would have thrown this paper away and never given it an initial thought. But if he is struggling to find a way to quit and he does want to quit then the book Nicotine Rage is the way to do it. The resources are excellent and when you read the journal, you can see that there are so many rough patches through the entire quit but that the end result is so peaceful. You just have to believe you can get there. And as far as about.com goes, nobody wants anything from you other than to help you out, and maybe you can help them out. Rarely do you see a gathering of people come together for no reason other than personal improvement and freedom from addiction of nicotine. It's an amazing, amazing place and I never would be where I am today if it wasn't for that site.

Please keep this letter and show it to him and if he doesn't quit now I hope he does very very quickly. If he doesn't believe me, have him go to the Hall Of Fame section on the web site quitsmokeless.org and read story after story of people that have quit smokeless. They are the same as him. It doesn't matter if he is a farmer or a professional athlete the Hall of Fame is full of people from every walk of life.

It's the single hardest and greatest thing he could ever do with his life. Trust me on this one. I know. And my kids/grandkids will know when they grow up too.

Rick in Tampa

Courtesy of Quitsmokeless.org

This letter can really change a person's life in an instant if you ask me. Yes he has a point in it can be the single hardest & greatest thing he could ever do with his life. It's the hardest thing I've had to go through in my 35 years of life. But I know that as every day passes, and that every day I glide farther and farther from nicotine in my body, I become better and better. Every morning I wake up feeling "slightly" more in control of my life, a little happier about things. I know that some days will come out of the blue where I feel all the symptoms again from the racing heart, to the depression, to anxiety out of nowhere. But I know it's just my body adjusting to life without tobacco.

Hopefully all you wives out there that are reading this understand not to take chewers personally and that it's the drug talking, not the person himself.

I know I was a real you know what when I would go on trips and not bring enough chew.This brings me to my next topic which I call Nicotine Rage!

Nicotine Rage

BOY DO I HAVE A STORY TO TELL YOU about Rich & I. We went on a cruise one year and thought it would be a bright idea to "go light" on the Kodiak. It was a week long cruise and he and I both figured we'd only bring 4 cans each. I remember discussing with him before the cruise about how we needed to chew less anyways so what better time to start then on a cruise right? Wrong!

Every night we were on the ship, we'd drink a lot of alcohol. Alcohol of course makes you want to use nicotine so for the first 3 days of the cruise it was cool because we had plenty of tobacco. But then by the fourth day, we had run out! We became very angry toward one another. We couldn't fatham to speak to one another because we both thought it was the other ones idea for not packing enough chew. Any little thing set the other one off. He would irritate me just by his tone and I'd irritate him by trying to teach him how to cope. So, one night it got so bad that we wanted to get in a fist fight over something lame. I'm glad both of us didn't because it wouldn't have resolved anything. We could have ended up throwing one another off the ship.

We'd managed to get off on the next port of call in Mexico that doesn't sell chew really.We must have gone through 8 different shops in Puerto Vallerta in search of it. We were determined. Finally we came to a place that sold it but it was only Skoal Straight, which wasn't the kind we normally did. It was nasty tasting and didn't really satisfy. We ended up having a poor time the 2nd half of the cruise because of no chew. When we got back to America we found ourselves at the nearest 7/11 buying Kodiak.

The moral of the story is, don't quit when you are heading into a potentially stressful or pleasant situation. It will just ruin things for you and whoever you go with. Quit when there isn't anything going on in your life and you can devote your time and energy to it only.

That was only one occasion of many. That one just stands out a lot because I can still remember those feelings and how fresh they still are. We both were on edge when we had realized we were empty on the chew and only in our 3rd day of a 7 day cruise. We suddenly went from enjoying ourselves to frightened and worried we wouldn't make the rest of the trip. I swear that chew was like a medication for us.

Quitting / Day 1–3

Ok so these next few days are vital for you. During the next 72 hours your body will be eliminating nicotine and a million other chemicals from your body that smoking and or chew injected into your blood stream.

The first day is hard because you are so used to putting something in your mouth at a certain time of the day. People will say to put a cough drop in your lip during that time but I just didn't do anything. I wanted to eliminate that idea that I was going to put something in my mouth at that time of the morning or day.

My mind threw a fit. I was extremely agitated, my mind wanted a chew! When I got home my wife was upset with me for not picking up after myself so I got upset and told her that she should be nicer to me since I was quitting nicotine. (I still ended up picking up after myself regardless of my tantrum.)

Eating food was good but if I'd eat too much then I'd want a chew after the meal. To me, it helped my stomach settle.

Quit Week 2

Based on what I've read and my own experiences with quitting on a couple occasions. People begin to complain that they can't sleep a full night or become constipated. Foggy or spacey feelings also kick into high gear. For me, the dizziness, spacey feeling was for real. I didn't panic though because I knew what a healthy guy I was and that this was all my body adjusting to the new me. Circuits were beginning to work together again on their own. In fact, some people that had these symptoms even said they kinda liked the feeling because it meant that they were healing.

Quit Weeks 3–6

Things get tough around this time because you are now in the beginning stages of your "Icky-3's" we like to call them. That stands for 3 days, 3 weeks, 3 months, and 3 years. Nearly all smokers or chewers have to go through some form of the icky 3's....For me, it was all of them. Especially 3 months. Be prepared to feel absolutely rotten and have every known nicotine symptom pop up out of nowhere. Its an uncomfortable feeling when you have a headache and are on the toilet with diarrhea but feel like you need to turn around and vomit and you are having anxiety because you think this is the way life will be from here on out for you. Yea, pretty crumby thoughts too. Those take a while to go away as well.

Month 3

I truly believe this is month through the 6th are the hardest part of any tobacco quit. Stomach problems for me were at their worst . Depression every single day that bruised my confidence and temporarily turned me into a homebody. Happy times turn into challenging times and it no longer seems fun but seems like a chore.

Like going out to a restaurant to me seemed like an effort because I'd get anxiety attacks sitting there waiting for my meal to come via waiter.

This is when I started getting crying fits out of the blue. I would feel hopeless all of a sudden and just explode into tears. People would call me at home like as soon as I'd be done with one of these fits and they'd say, "Oh do you have a cold or something?" I'd try to tell them yes, but they could tell then it was more then a cold… It was depression in high gear.

Crying made me feel good even though I didn't enjoy actually doing it because I couldn't explain what the reason for it. People cry when loved ones pass away, or a divorce, or something they've lost but not just for no reason.

Symptoms that nag

Sleeplessness is a bad one because it makes all your other symptoms worse depending on how much you get.

Insomnia was and still is a tough one for me now that I'm quit. A large percentage people that quit chew deal with this through medication. I chose not to and just deal with it by taking vitamins at least 3 times a week and drinking water.

Naussea/IBS is common when you quit and it was a primary problem for me in months 2 to 6 (you'll be reading my day to day journal later in the book so that you can understand what you may have to go through too). I'd wake up in the morning to this strange morning sickness. Often times it would wake me before my alarm clock and force me onto the toilet. After 6 months, it should start getting better but it can hang around on and off.

If you suffer from IBS and quit tobacco, be prepared for a lot of bad nights. Your IBS symptoms will get bad. You may experience episodes that last 5–20 days even. Don't worry, things pass over time but you'll have to be patient or take medication (which I'm not a fan of.)

Depression bouts will happen a lot, especially if your sleeping pattern isn't solid. I found that on the days that I slept well, I had very little depression but on days that I slept poorly, I had to fight myself to think positive. I would try to remind myself to not take things too seriously on these days. Just survive life right now and live life later. Depression is a real serious thing and you can feel like you are in a prison of your own mind at times even though you are free to run outside or take a trip.

Depression makes your life so that when you wake in the morning, you have a dark blanket over you. You don't want to speak to anyone and it hurts to even see a person disappointed in you.

For me it was seeing others close to me upset because I couldn't do all the things they wanted me to do. They'd feel I needed to get on medication. People would try to convince me to take them but I knew that if I did go that route, I probably wouldn't be alive writing this book.

Only I knew how medication made me feel and it made me suicidal. I was better off not putting this stuff in my body. So it's important that you not listen to anyone else but yourself about that. If you are low….GO AHEAD and be low. It's okay to feel down; it's not the end of the world. Like Jeremy Kernion said, "You are not the center of the Milky Way galaxy" Things do not revolve just around you.

Once you are able to not feel bad about not doing anything or being an introvert, people will learn to accept your NO answers. If you were prone to being involved in every event out there, it's not the end of the world if you take a year off from all that and hide from major social events.

When it comes to depression, you have to fight as if you are heading off to war! I mean REALLY fight!

Here is an article Michelle Boisvert wrote on about.com regarding Depression:

Depression is an issue of focus for many who quit smoking or chewing tobacco. At times, quitting smoking or chew is an extremely difficult thing to do, and it is difficult enough when you're feeling happy. I would venture to guess that many people have relapsed during periods of depression; it is more difficult to stay focused and maintain resolve and motivation. For those people who suffer bouts of depression during the cessation process, the condition is usually mild and temporary.

People who have been diagnosed and or treated for depression or anxiety prior to quitting smoking or chew should monitor any changes in symptoms carefully as they begin and move forward in their smoke-free / chew free journey. Some may experience changes in their symptoms, which may necessitate a change in treatment. If you experience drastic mood changes when you quit smoking or chew, or if they persist for an extended period of time, a visit with your physician is advisable.

If you find yourself suffering the more common temporary emotional upset, try to relax and let the feelings come as they will. (This is all too common that these odd sensations will creep up out of the blue since you are used to your emotions being stimulated by the nicotine.) Quitting is a big change in lifestyle, and you will react, to some degree, both emotionally and physically. And think about it; there really is no reason to deny your emotions; they are personal, and they belong to you.

Courtesy of About.com Smoking Cessation Forum

Symptoms of depression may include:

- sleeplessness
- sadness
- difficulty concentrating
- anxiety or an empty feeling
- fatigue
- changes in appetite (more or less)
- loss of interest in hobbies, activities
- emotional irritability
- and many many other symptoms

Now here is a list provided to me by anxietycentre.com. It's a list of "most" anxiety symptoms. When/if you get some strange symptoms starting in after your quit, check this list and see if its on there.

Please note that these symptoms will be experienced at some point during your quit. Not all but a large number depending on how much you chewed tobacco. I would be surprised if you didn't get anxiety at some point in your quit.

Anxiety Symptoms

Anxiety symptoms are no joke that's for sure and when you get them, they can really throw your whole life out of whack. I know that they changed my whole outlook on life. The key is to remember that these sensations will not kill you. You'll have to learn to live with them for a while during your quit of tobacco. I've had about 90 percent of these symptoms I'm listing below and about 80 percent of them came after quitting chew so be prepared and don't say I didn't warn you this would happen....You are quitting a substance that was hiding all these sensations, thoughts and emotions. Here you go:

- Blushing
- Allergy Problems
- plugged up nose
- Body Jolts or like an electric jolt sensation
- Your skin burns or feels dry
- Chest is tight for no reason
- Chest pain
- Choking sensation when you eat.
- You think you are going to choke when you put food in your mouth

- Fatigue, exhaustion or the feeling you need to sleep NOW
- Clumsiness or you drop stuff more often or trip and fall more.
- Cold or hot body
- Crave salty or sugary foods
- Difficulty talking or speaking when spoken to and you get tongue twisted easier
- Feel different, odd or wrong
- Flu symptoms, feel ill
- Heart Palpitations
- Hyperactivity, nervous energy
- Buzzing in legs like a bumble bee is buzzing around
- Decreased sex drive
- Muscles that vibrate or shake
- Muscles twitch
- Nausea
- Throwing up (occasionally)
- Stiff neck
- Waking up hot and anxious
- No energy, tired
- Numbness & tingling sensations
- Constant Muscle stiffness
- Heart beats too hard
- Ribcage tightness, tight band feeling around ribs
- Shooting pains around neck and face
- Skipped heart beat
- Startled super easily (this has happened to me a lot)
- Floor feels like it is moving down or up for no reason

- TMJ
- Trembling or shaking
- Dizzy, lightheaded
- Constantly having to urinate
- Feel weak, low energy
- Weight Loss or gain
- General concern about your heart
- Feel like you have to think about each breath you take
- Feel smothered
- Shortness in breath
- Constant yawning to try to catch breath

Now for the emotions and anxiety

- fear of what people think
- afraid to go somewhere and being trapped
- Feel overwhelmed
- Avoid public
- Afraid of dying
- Afraid of losing control
- Impending dreadful doom
- Afraid of passing out
- Losing your mind feeling
- Irrational thinking
- Heightened self-consciousness
- Have to find nearest bathroom before you can relax
- Always need to sit near exits
- Afraid of everything
- Altered state of reality
- Derealization
- Spaced out feelings
- Trapped inside your own mind feeling

- Bad dreams
- Tickle in ear like you have earwax

Mood and anxiety

- depression
- keyed up feeling
- sad
- feeling of wanting to cry for no reason
- sad feeling of thinking your loved ones will die soon
- emotions feel wrong
- dramatic mood shifts
- insomnia, up thinking about life
- have no feelings about things you used to enjoy

Sight and Touch / Anxiety Symptoms

- Dry eyes that get red
- Eye tricks like you see something at the corner of your eye
- Sensitive to light
- Your depth perception feels wrong
- Bruning skin sensations, skin sensitivity like you haven't used enough lotion
- general Pain
- Pins and needles sensations

There's probably even more that I can list but I hope I got the big ones out there.

While quitting chewing, the body and the mind are in a state of transition, and it's not uncommon for new ex-chewers to struggle with their emotions.

It helps to build up a list of of gratitude. When you're feeling depressed and having a difficult time finding the positives, quitting chewing is always there; you can always feel good about that. In other words, if you feel like trash during this quit, you can at least have the quit chew part to talk about as a positive in your life. Take time at the end of the day to acknowledge the importance of what you are working to accomplish. It will help you fill that list out.

Find comfort in knowing that you are being kind to your body. Gratitude will help get rid of negative feelings and make it easier for you to manage depression due to quitting tobacco.

Change Your Mind

One of the greatest challenges new ex-chewers face is a very important change in beliefs. It is that shift in thinking from seeing chewing as an exercise in deprivation to realizing that it is in fact one of the best gifts you've ever given to yourself. This is a crucial step in the process, and it is with this transformation that many see the symptoms of depression begin to lift.

While you are moving through this transitional period, crying, and even screaming are preferable to putting chew in your mouth. Early in the quitting process, some resort to chewing because they perceive chew to be something that offers comfort.

Chew is not comforting; it is familiar

Time away from smokeless tobacco makes it less familiar; you'll begin to see nicotine for what it is...a highly addictive drug which robs people of time with those they love and seeks to kill, one dip at a time...if you let it. Give yourself time to develop new and healthier coping strategies. Don't buy the lie. There is never a good reason to open up a can.

It's likely that many of you who are feeling down, felt happy recently. You'll feel that way again. In the meantime, you can find comfort in turning to your friends, your family, or your faith. In time, these will become the more familiar sources of comfort to you, and chew will become that thing that you thought used to make you feel better. Take comfort too, in knowing that millions of people have been through this process and not only have they lived to tell about it, but many include it among the most rewarding and life-affirming experiences of their lives.

So, if you've quit chew in the last few weeks or months and are feeling blue, please be patient with your mind. You are going through some big changes, and your body and mind need some time to adjust and heal. Do any little things you can think of to give yourself a little lift. Rent a comedy, try a new hairstyle, call an old friend, go to the dollar store and by cheap their cheap toilet paper, get a facial, find a picture of a person you want to look like,spend a little of the money you're saving since you quit, and buy yourself something special. Keep reminding yourself that it is a temporary state, and you'll find happier days ahead; with them will come a tremendous sense of pride and empowerment.

Traveling

Each person out there is different in how their minds respond to nicotine withdrawal. For me, I had a difficult time traveling very far from home. I'd try to get away and take off for the weekend but would end up at a hotel sick and depressed as a dog. I wanted to be close to home, my element, my starting point.

It got to the point where the only thing I wanted to do was just do a certain number of things, go grocery shopping, walk through town, church, work, and home writing, reading or watching TV.

These are usually your turning points with the anxiety and depression because the more you fight it with this uncomfortable feeling, the more you'll get through it. I believe it was Winston Churchill that once said, "If you are going through hell, keep going!" I can't agree with him more. What else are you going to do? Turn back and start using tobacco again?

Chronic fatigue Syndrome

Chronic fatigue is a common one for those getting off smoking and chew. You can have a solid nights sleep and still find yourself with extreme exhaustion followed by depression from all the confusion you may have as to why you are experiencing this.

It took me a good 15 months off of chew cold turkey to get rid of the depression ½ but the fatigue stayed with me. It hits at various times of the day when you'd normally have been pulling out your can of Copenhagen. Now that those times aren't filled with a chew in your lip that would normally provide you your caffeine (yes that's right caffeine is in chewing tobacco also) you won't have that boost to pick you up for the remainder of the day. It's the pits yes, but it does eventually get better.

You learn to cope with this drowsy feeling. When you first start experiencing it in the beginning of your quit, it can be very scary and confusing cause you think something is wrong with you. This is all part of the game though.

I did a post on a quit tobacco forum to see if any others experienced this.

Here is what I put:

I've been noticing that I get extremely tired during the day now. I feel like it will hit all of a sudden. Following that, the stomach will get the nausea followed by gas. I haven't been sleeping well but this is far worse of a feeling then just being tired.

And the Responses were:

Been there, done that. For me it was more like my brain shutting down and was absolutely frightening when it began happening every afternoon. Since supplementing with vitamins, minerals and anti-oxidants has stopped that for some 10 years, it seems pretty obvious that there is something about this condition that can really affect normal brain function.

Another poster:

You are not alone—this has recently been happening to me too (say, around a month?). By and large I would say that I've been feeling all of the nausea so far. Anyway, I just wanted to let you know you are not alone. Apparently, Chronic Fatigue, is commonly associated with IBS and quitting nicotine. I am due to go and see my doctor within the next week or two, so I shall find out more then. (He went to his Dr. and got back to me about this, and the Dr. explained that this was an all too common issue that he sees in his patients and said that the best remedy to combat Fatigue is to....You guessed it — rest!

Another person replied:

I am battling this to see if it is a body-demanding-sleep thing, or just me wasting my life. I have now completed 4 days without my mid-day nap, and I am totally cream crakered! I'm so grumpy it's almost incredible, and I feel so tired that I'm having a job staying awake at 10 in the morning. I'm glad I'm not driving because I feel I would be a danger to everyone. I need help!

Responder:

I had the same problem. I never went to Drs. and then all of a sudden I was sure there was something wrong with me. I had Irritable Bowel syndrome and awful fatigue and just a feeling of overall garbage. I felt like i had been drinking Nyquil all day. I have now been free of IBS symptoms for about two years now and my fatigue is gone. One thing though, maybe get a physical...under active thyroid can make you extremely tired (I had gotten my physical as he stated to do and everything was fine.)

I couldn't tell why I was so tired all the time. I started to notice that I felt better when I didn't eat and I almost always felt more alert at night (after not eating for a while). It wasn't until we took a 4 hour car ride and then stopped for Chinese food. I rarely eat Chinese, but shortly afterwards I felt so tired, and cranky and unwell. My IBS was awful as well. When we got home I looked it up on the internet and sure enough it is the Free Glutamate in foods that gets me....also known in some cases as MSG, but even foods labeled "No MSG" can have them. There were days the only thing I would have was coffee with a little half and half so I was convinced initially maybe I had a milk issue, but when I checked the label there is something in it called Carageenan (free glutamate). Now I drink coffee with organic half and half (no additives) and I stick to Atkins diet type foods...things that can't be altered by man such as meat (some has it though) veggies, fresh fruit etc. I can now handle it in very small doses, but the way I feel afterward I can tell you there was a suspect ingredient in it just by how I feel afterward. I have my life back again and it wouldn't surprise me if it isn't affecting some others with IBS as well. If you feel bad after Chinese food, there is a great chance it may be affecting you on a daily basis in the hidden name of "soy protein" "spices" "Natural flavors" etc.

Go to http://www.msgmyth.com/ Hidden names for MSG and also check out the symptoms. I finally found out what was wrong with me after suffering IBS for 5+ years.

Best wishes. Hope you get to the bottom of it,
Kari

People went on to say that MSG causes problems with the stomach which causes more problems with your fatigue. This was posted in an IBS forum because of the fact that a large number of people on this board had recently quit smoking or chewing and started experiencing all sorts of stomach issues from nausea to abdominal cramps.

It just goes to show how tobacco can affect every single part of your body. Who would have thought it would cause a person stomach aches? A lot of people that responded tried to blame their constant nausea on the foods they ate (as you read in Kari's case above) or how much they consumed but I'm not buying it. It's from the tobacco and the withdrawals.

One person went as far as to say that anytime he ate anything with flavor, he would get sick so he'd just eat plain food all day. Poor guy! If he only knew that over time, he'll be able to consume foods as he did before he used nicotine.

When I made my decision to quit chew I never thought it was going to be this crazy and that I would experience the kinds of things you will read in my quit journal. I believe my mind went though every emotion that it can possibly go through during an eight month period. I'd get really low for five minutes followed by a feeling like I'm all better now and everything is all good. Like a tidal wave that keeps repeating itself. Up and down, up and down.

Read how when I'd have these crazy emotions that I would start to forget that it's all a part of nicotine withdrawal. I never thought that I would turn from normal shy guy to nutcase in a matter of days. People that previously talked to me normal would talk to me with dignity and concern now as if I had a mental disability. Before I quit I was thinking I'd be fine within a couple weeks and would be strong and confident again. *No*, that wasn't the case.

My Quit Journal

OKAY THIS IS WHAT I DECIDED TO DO. Everyday that I quit was another day that I would log down my thoughts and feelings. Try to notice how on some days I was really losing my mind and started to think there was something really and truly wrong with me mentally. It's important for those of you quitting chew to read this section over and over again until you understand this experience.

You are quitting for the long term so this experience in the short term will be well worth it. You will feel the most horrible you've ever felt in your entire life during these next six to eight months after quitting but these are the times to work on yourself.

I bring to you my personal journal and will be letting out to the public my most private moments during this quit. And although some of this you may think does not pertain to quitting chewing tobacco or nicotine, it all relates. I never before experienced this kind of drama before. I was normal, able to function and do things like normal human beings. The crazy thing as you read my journal is, the real head games the withdrawal plays on me didn't come until around three months after I quit. The first and second month turned out to be a cake walk (as my father in law likes to say about teams that play his UCONN basketball teams).

Hopefully you'll take the time to read this and not skip over it. You can keep track of your quit, through my journal and realize that you don't need to be admitted to an insane asylum.

7/10/08 day 3 off chew

Everyday I'm feeling better and thinking clearer. The first day was difficult for me, I felt sick to my stomach, dizzy, floating feeling, short breath, anxiety, depressed, and couldn't sleep. The next day I slept a little better, then last night. I probably slept 8 hours! Today, I had some anxiety at Subway Sandwiches and again in the HR Directors office when discussing hourly wages.

7/15/08, 9 p.m., Tuesday/8 days off chew

Today made it 8 days since I've had a chew. This hasn't been easy because of having to replace the time that I'd normally be chewing with something else. I've been trying to occupy that time with walking, reading and working out. Still get dizzy when standing and talking at work but not when seated. The thing I learned is one day could be a non-craving rather easy going day while the next can be extremely difficult. Today I've been thinking a lot about chew but I know if I was to have just one that I'd be right back to where I started at 8 days ago.

7/16/08 7:30 p.m. 9 days off nicotine

I've been feeling very low today. I began feeling this way right around 5:15 p.m. on my drive home. This has been my toughest day it feels like. I didn't eat particularly healthy either. Today I had chicken, watermelon, water, Nutra grain bar, burrito that had hot peppers and cheese in it. Then tonight I had steak and macaroni. Depression has been the major difficultly this evening.

7/17/08 10 days

The temptation to chew is growing. I tend to try to find ways to escape the situation. I get frustrated with any little thing that a person says to me it seems. My temper is high.

(Now notice how personal I was beginning to take things at this point in my quit. Everything seemed to make me irritable.)

7/22/08, 7:15 p.m., 2 weeks off chew

I am not feeling that craving anymore really but instead I feel defeated by the stuff. It has conquered me in every way possible. Paranoia, loss of sleep, sick to my stomach, irritated, dizzy, very strong depression, anxiety when being around people that want to socialize.

My appetite went away these last 48 hours and have noticed a new stomach ache that starts every morning. It's like a bee buzzing around in my gut. I remember this from my first quit too.

7/27/08 Day 19 — no chew

These last 3 days have been anxiety filled. Yesterday at the Sardine Factory which is a fancy restaurant, I needed to leave fast because of a panic attack in there. When I originally sat down at the table I felt confined for some reason. I got hot and stuffy and even a screw driver drink wasn't helping much. It seems like every time I'm at a restaurant I feel this surge in my chest. I don't get it at home or in non-social situations though.

7/28/08

Felt good today, no depression, some dizziness & tiredness. Foods I ate today were pineapple, turkey & cheese sandwich, nutra grain bar, butterfingers, tea and water. Mentally though, it felt good to feel normal again.

No cravings for chew

7/29/08 Day 21

Today was okay until around 3 p.m.. I suddenly became extremely sleepy. I started my day with 3 chocolate cookies followed by a stuffed bell pepper my wife made the night before. I was over full by 11 a.m.. When I got that way I'd start thinking about it and that's where I could feel depression was looming. I didn't sleep well last night and woke up at 5 a.m.. I have mild cravings for chew.

7/30/08

Today was a great day until I got back from lunch. I had eaten sushi for lunch but I believe I overate again. I became extremely tired and depressed. Negative thoughts came into play. I'm wondering why I'm not starting to feel better from being off the chew so long. It's been 22 days cold turkey now.The Foods I've eaten today—2 cookies, tea, peanut butter & jelly sandwich, sushi, catfish. Depression stuck around until I got home and we went for a walk at 7:30 p.m. Something about going for a walk really clears my mind from depression, not all the time, but a large percentage of the time. I know during the day I kept saying to myself "Why am I feeling this way? There is no reason to be bummed out now." I am very frustrated, here I should be feeling good from my quit but instead, I'm going through hell. Mild cravings for chew

7/31/2008

"Stop day dreaming and face your problems"—*Neverending Story*. Today was the best day I've had in over a week or so. I believe my mind is finally starting to calm. Even though I got up twice last night to go to the bathroom, my mind woke up fresh. Very minimal anxiety & depression today and when it came, it would only last a moment. I made sure to eat small amounts of food and to not get too stuffed. It keeps me from getting tired (note how confident I became when I was able to get a goodnights rest. Sleep really proves to be a deciding factor in a lot of peoples quits because it keeps you thinking correctly.)I have thoughts of chew but mild.

8/3/08

It has been quite a few days now and I'm beginning to lose track on how long ago I quit chew. One thing I can say is that I'm a lot more focused on things then I used to be before the chew habit. Sure I'm gaining a lot more body fat that I don't like but I'd rather have that then anxiety, depression and the stomach aches I used to feel just after chew. There were endless days that I'd chew too late and not be able to sleep at night which would cause depression and fatigue. I'm thinking of beginning on another book. I know that it will require dedication and time, both of which I have plenty of now. I have not figured out a subject but I do want it to be something where I can help others with their problems. Possibly about addiction to chew. Maybe I can help a few guys out from all this pain I'm going through. Thoughts of chew throughout the day.

8/4/08

I got in late to work today, but I had no signs of depression which I found refreshing considering I didn't sleep much last night. I only got touches of depression when I'd stop and think about things and take inventory. (Notice how I'd have these good days and how good I felt. Good days were much appreciated since it seemed like everyday felt horrible.)

8/5/08

I still have irritability and a short fuse. Like I'm getting annoyed with my boss asking me to go do scanning which normally I wouldn't take personally but today I feel like he's nagging me. I can tell that this stuff is getting to me and I could use a chew right now. Seems like the only thing that would get me though this stressful time. Some mild thoughts of chew

8/13/08

Today was fine until 4 p.m. I became a little depressed that lasted a couple hours. I woke up again to a stomach ache in the morning. Could chewing tobacco help me out?

8/26/08

I woke up tired this morning. Woke up early and couldn't fall back asleep. Today was very difficult and the toughest I've had in quite a while. Last night I smoked a cigarette and had 2 drinks at dinner. After I smoked I felt a major buzz but knew I'd get sick from it and sure enough, I did. Right now I can't have anything it seems. No alcohol or nicotine.

8/27/2008

Today was another difficult day. It has been 48 hours since I smoked that cigarette. It takes 72 hours for nicotine to be completely out of the blood stream. Today I felt a loss of appetite, depression beginning around lunch time. It hung around until about 6:30 p.m. I layed down and closed my eyes for 45 minutes when I got home from work. Funny thing is I noticed my anxiety and depression have a habit of vanishing around 8 p.m. every night or when I go for a walk. I learned today that when I am tired & depressed it does no good really to go for a walk. I must first rest when I'm tired, then walk when I've gotten my second wind. No thoughts of chew

8/29/08

It has been nearly 8 weeks since I've chewed even though on Monday (3 days ago) I had a cigarette. Now that I don't use nicotine, when I see people using drugs or alcohol, it annoys me. I watch the television show "Intervention" and see how some of these people get when they need their fix. I was the same way when I was chewing so I can't get too upset with them. But at least I know now what an annoying person I was when I didn't have my chew.

I felt good today and didn't feel super tired until 5:30 p.m. or so. Made it farther then usual this week. Had depression and tiredness from 6:45 to 7:45 p.m., during that time I laid down and closed my eyes.

9/2/08, Week 8

It has now been 8 weeks since I've had a chew. The last couple days have been excellent and I've slept.

Today I was tired at work but it wasn't the depression kind that I'd been feeling leading up to today. It was more the exhaustion, rundown feeling you go through when you've been up more then 16 hours in a row.

9/9/08, Week 9

It has now been 9 weeks that I've been totally clean from chew. I've noticed that I don't journal as much as in the beginning. I believe the reason is because I'm getting better. The only time I ever write about it is when I'm having issues. Like today I've been having problems on & off since this morning. Stomach problems, tired, depressed feelings. No more of the anxiety stuff though which is nice. Today I've had a lot of gas and indigestion though. Then again, yesterday I ate clam chowder, cheese and other gassy stuff. It's hard to describe what I go through in my mind when I have these moments but I can say, its not easy. And it sucks when you suddenly don't look forward to your vacation and are instead worried about getting sick far from home. Some thoughts of chew

9/11/08

Today was very blah. When I got done working out at lunch time, I felt a sudden tiredness and depression. I've noticed a pattern that whenever I get tired I wonder when depression will follow. As soon as I do that, depression will strike! Thank God I have a job and things to keep me busy during these challenging times. I never thought quitting chew would cause this many problems in my head.

Saturday, 9/13/08, 1 p.m.

Woke up at 7:45 a.m. and had a little stomach ache and felt some depression. Not sure why but lately I've been waking up before I should. It's hard for me to sleep in nowadays. The last three days though I've woken up earlier then I should be. I'm hoping it's all just a phase to this whole quitting chew thing. My Uncle Hugh seems to think so even this far out in my quit. Thought about chew but didn't want one. Think my life would be different if I started in again though.

Saturday, 9/27/08, 5:45 p.m.

Woke up extremely early this morning and my back was all hot. Went and slept on the couch. I ended up sleeping in until 10 a.m. I was proud of myself because lately I have had insomnia. I've been extremely sick to my stomach too. I know its all because of getting off chew and my brain is still going through weird things that cause all this. Three days ago my friend Dave West and I went to Chevy's and had a couple Margaritas. I ended up getting very sick for the next one and a half days. I emailed my Doctor and he said I could have intolerance to alcohol. I agree with him. Today, my body is still severely tired, but I've discovered when I eat a couple bites of smoked salmon, it's very uplifting. My spirits get high within 10 minutes of consuming. I'm still also getting over what I believe is the flu which had begun 6 days ago. I haven't had much of an appetite.

Sunday, 9/28/08, 3:10 p.m.

Today has been another difficult day for me. I woke up early to that barking dog next door and couldn't fall back asleep. Before I quit chew, I'd sleep right through the barking! I'm going through the most difficult time of my life. I don't understand why I'm having all these problems either. I have to remind myself that it has to do with quitting nicotine and nothing more. I'm not on any meds nor am I hung over. Symptoms are loss of appetite, sad for no reason, awake thinking at night. I had to leave my parents place after an hour because I couldn't take all the noise and yelling. I feel weird "vibrations or buzzing" coming from the insides of my legs or thighs. When I feel this in the morning, I know that I'm in for a long day filled with anxiety.

That vibration always is a sign that I'm keyed up. The funny thing is I never used to be like this before when I was chewing tobacco. Maybe I should get back on it. Although with chew I'd have a different sort of problem. I'd get sick 5 minutes after putting a chew in, anxiety started in towards the end there, I'd flake out on doing things, my heart would have a sharp pain, I'd be short of breath (both of those I don't get anymore now that I quit.) Right now is not the time to be planning anymore stuff. It's just too much food on the plate to be concerned about traveling around. It wouldn't be fun "getting away."

9/28/08 7:15 p.m.

I just got back from a long walk with my wife. During the walk I felt good and once we got home I laid right down on the couch. No sleep but closed eyes a little. Was that stomach pain that started in again? I know that I just had my physical and Dr. Gorman insists it's caused from stress he said. Ate some avocado dip and the pain went away. My wife thinks I should get on anxiety pills but I don't want to. In my last quit 3 years ago, I tried taking a number of different ones and they all made me even more nuts then I already was! I won't name them out because each person responds to meds differently.

So I knew how pills made me. I know once I get a good night's sleep I will be normal. I know how my body operates better then anyone on earth. I prayed a lot today. I also pray that my prayers would be answered a lot faster! No thoughts of chew

9/30/08

I have been getting irritated easily today. The noise from the TV was bothering me so I turned it off but my wife didn't want that sound off so I put in some earplugs. I need more quiet time in my life, away from noise, people and cars. California doesn't have a lot of space anymore. And yet, my mind won't allow me to go to a faraway vacation. What's up with that?

Today I've had diarrhea, sick to stomach like the flu, depression, and also fatigue. I would go to bed now but it's too early. Have to wait until at least 9:45 p.m. Today I was fine until about 2:00 p.m. when I started taking inventory on myself again. As soon as I did that, my problems began. I was reading a discussion group about nicotine withdrawals and a few were just like me with the upset stomach, fatigue, insomnia, anxiety and depression. I try to remember if I was this way too at my last quit when I had to move in with my parents it got so bad. And the answer is, yes, it was just like this over there. I would get depressed and go out back where the chicken coop was and hang out until dark digging up worms for the chickens.

10/01/08

Today was another rough day. I had to force feed myself. I've been sick to my stomach through out the day and having major fatigue that is causing me depression and have sad thoughts for no reason. I know that people in the about.com smoking cessation group keep reminding me that this is common to feel this way but I just don't get it.

I'm beginning to think I need meds now. Insomnia, depression, anxiety, nausea are the main problems. It feels like the flu but it can't be because I've had my flu shot. If I still chewed or started back in I wouldn't be having these problems. Could this all really be a result of quitting chew? Could my body and mind still possibly be reacting to not having the drug in my blood stream anymore?

10/1/08, 9:30 p.m., con'd

Times get tough in an instant though. One minute I'm fine and the next I'm focusing on the moment and impending doom is lurking. I'd get worried about one thing or panic when I couldn't fall back asleep at night. Besides that I'd get sudden stomach nausea and would panic that this feeling is going to stay with me wherever I go. It's not a cheerful thought and just making matters worse when I think about it all too much. Maybe I should consider using tobacco again to get my brain normal. But if I do that, then I lose my quit. I better not go that route.

10/2/08, 9:20 p.m.

Today turned out being a good day even though my visit to Dr.Gorman's was tough on my stomach. My heart rate was 83 and blood pressure 131/79. I was really keyed up sitting in the wait room for my name to get called. At work today I was very tired most of the afternoon up until around 6 p.m. When I got home Dr. Roger Johnson, a therapist, called me. Told me how I need to learn not to concentrate on my problems too much but instead focus on something like the baseball playoffs. I need something to "redirect" my attention from the symptoms. I did as he instructed and the symptoms wore off for the rest of the evening.

10/03/08, 7:00 p.m.

Went and visited my therapist Dr. Johnson today. If only had one bad bout today with fatigue and depression.I had an okay day today. I no longer crave tobacco and every day that I feel so rotten like recently, I regret that I messed with chew more and more. It's like I should have known the very first time I tried it when it made me sick as a dog.

10/4/08, Saturday, 12:30 p.m.

Woke up at 7:15 a.m. and had more diarrhea until everything I ate the night before was emptied from my system. Did my heart rate today and it was 135/65 with a pulse of 67. I have that keyed up feeing in my legs again still and it's like a trembling that won't go away. I would like to calm down but can't on my own. There is no reason for this hyper feeling though. I have nothing going on that I should be nervous about!

10/05/08, 9:00 p.m.

For some reason I snapped out of it last night. Sheryl and I ate salmon and stuffing with gravy followed up with Rice Krispy Treats for dessert. This hit the spot. I had a full nights rest and woke up feeling a lot better. As today has progressed on, I have eaten more and more. Started the day with and English muffin, and after church I had Spam & eggs. I later went for a walk with my wife and we had a salad when we got back. Came home, read for another hour, had some more Rice Krispy treats, then went for another short walk. Then for dinner, had pasta, sourdough muffin, and some canned veggie soup. My mood right now is stable and thinking is normal once again. I believe I just went through what they call “The Icky 3’s” which is also known as 3 days, 3 weeks, 3 months and 3 years. The idea is that on or around those times in your nicotine quit you experience discomfort for long periods of time.

10/6/08, Monday, 7:00 p.m.

I was able to get a decent night’s sleep and was able to eat a sandwich, chicken thigh, and the remaining few bites of salmon. Today at lunch, I began feeling fatigue followed instantly by depression. This lasted mildly until around 3 p.m. Then when I got back to work I had 2 tacos then at home, had fish dip that my wife buys. After about 30 minutes my stomach felt funny so I took a pepcid called Famotidine 40mg which Dr.Gorman had prescribed for upset stomach. After about 20 minutes I got sick and vomited a little. I was really sick. But as fast as it came, it went! I just couldn’t figure that out. I had heartburn all day today too which is a new thing! What’s next to this quit that I haven’t dealt with!?!

Wednesday, October 8, 2008, 8:30 p.m.

Today was a good day all the way until around 7:00 p.m. I felt fatigue and a little depression. I wondered if I could handle being around my family this weekend. I started thinking I’d have issues again. It’s funny because I have been doing well the last 2 days. My symptoms have been sadness, fatigue, tired, and depression. I have not had that stomach pain for quite a few days now which is nice. I'm thinking it may have passed. (Note how symptoms will come on for a while then vanish suddenly.) No thoughts of chew through whole day.

Thursday, October 9, 2008, 9:00 p.m.

Today was a difficult day. Last night I was woken up a couple times because of stomach issues and having to urinate constantly. I went to the Doctors for a physical and it turned out good. He checked my stomach, heart, ears, throat, thyroid, and eyes. He did blood work as well since I told him how cancer runs in the family. I'm sure I'll get to see the results very soon. Today has been another challenging day dealing with the minor stomach gas, lunchtime depression, fatigue and moments of bad, racing thoughts. I know I can make it through this with patience and remaining calm when an unpleasant thought enters in. I am happy though that I went through with the blood test and didn't wait on it. I am not a fan of having blood extracted from my arms. (Note—2 days later the test results came back and everything turned out normal.)

Saturday, October 11, 2008, 12:30 p.m.

I did my pulse & blood pressure at Long's and it came out 114/68 with a pulse of 98. It's down from the Doctors office where my pulse was 104. Still a little on the hyper side today.I'm trying to focus on slowing my breathing down. I sure hope this is all nicotine withdrawals because if this is how my life is going to be forever, forget it. I'm out of here.

Monday, October 13, 2008, 9 a.m.

Sometimes I wonder if it was worth it for me to quit chew. I mean its not like I am better off now then I was when I was using nicotine. Now I've got all these problems it seems like. The depression is making all this stuff difficult on me. I don't like feeling hopeless.

I try to picture myself going on a vacation but my mind just won't allow the good thoughts to last long. I know I was reading somewhere that if I can get this sad that also means my mind has the ability to get that extreme happy too. How to do it and conquer the mind is a whole different story. Its going to take time, a whole lot of precious time. Its going to take patience and time. No thoughts about chew

Wednesday. October 15, 2008, 9:25 p.m.

Last night I took a shower before bed and that turned out to be a mistake because it woke me up. Today was hard being at work because of the fatigue. I'm wondering if I have chronic fatigue syndrome. I can say on a positive note that I'm not getting that deep sadness like I was getting before. I've noticed too that everything sounds louder and more crazy and in my face. I've discovered that when you have a tired mind, it's hard to even think straight. Today makes 14 weeks or 98 days. I slept great this weekend and I felt good for the most part so it goes to show that if I sleep, things are easier and viewed more realistically then when I'm dealing with exhaustion.

Tuesday, October 21, 2008, 8:00 p.m.

Tomorrow will be my 35th birthday. Today is my last time in my life that I will see 34. Lately I've been getting constipation and still some trouble sleeping at night. I have difficulty staying asleep at night. Getting better though and depression is under control (a little bit.)

Wednesday, October 22, 2008, 2:15 p.m.

Woke up at 6:50 a.m., stomach wasn't feeling right. I've had diarrhea all morning long. Went to Burke Williams and cancelled the appointment after sitting in their hot tub for 15 minutes. I wasn't feeling right at all and my being sick made me nervous. Last night I ate 3 spicy chicken wings, 1 muffin, artichoke dip with mayo and water. I used Garlic chips to dip into artichoke sampler. Earlier in the day I ate a cup of noodles, 2 peanut butter and jelly sandwiches, 1 peach tea, and an orange. Reason I explain all this is because my stomach just doesn't like anything. And today has not been fun. I also slept bad last night. I don't know if its nicotine withdrawal that is still causing this or an actual sickness. It has to be one or the other or even IBS.

Today I got really sad and began crying for no reason. I couldn't understand why but it wasn't fun. I know that when I get sick with the flu, my anxiety, depression get bad because I start thinking this is just the way it is and will always be and that's not the case. I have to constantly remind myself of how others are dealing with their quits too. In about.com most of them deal with the symptoms by getting on medication. This is a real

Saturday, October 25, 2008, 6:08 p.m.

Today has been a day of fatigue and anxiety. I woke up 3 times last night so I didn't sleep solid. I want to sleep better tonight so I will stay awake until 10 p.m. at least. Last night after watching a movie involving people jumping off the Golden Gate Bridge, I got tired and slept on the couch.

This morning I had stomach problems that began at 6:30 a.m. I emptied my system and didn't feel comfortable until 8 a.m. and that's when I was able to actually sleep again. I slept until 10 a.m., took a shower and we went to breakfast. I had anxiety when we sat at the dinner table. Felt like my pulse was extremely high. When we left the coffee shop though I instantly felt better and regained my appetite. See, when I eat at places that involve waiters serving us I get this anxiety thing (not all places do this to me like Marie Calendars doesn't). But when we eat at places that don't require service where you just order and pick up the food at the counter, I'm fine sitting in the restaurant eating. I know this stuff isn't going to last forever, it just can't. I know I have to fight this and know that its just part of nicotine withdrawal.

Sunday, October 26, 10:15 a.m.

Still don't have my appetite back yet and can still feel the energy in my chest. My thoughts are still not 100 percent yet either. It seems like I don't know how to handle myself when these emotions come in and out. Like they come through and beat up my mind and body and then vanish into thin air. I'll feel fine for a while then something negative follows. I would say its 60/40 that the amount of time I'm awake is good compared to the bad moments.

It has been a mixed bag because I didn't sleep enough last night. I'm paying for it now. Made it to the gym today and had an average workout. When I got done is when the tiredness set in. When I woke up this morning I had to hit the snooze button a few times. I didn't fall asleep until around 1 a.m. and woke at around 5 a.m. But like Jeremy Kernion says, "No one ever died from being tired." It's been challenging though with these moments I'd had. These waves will come from nowhere. This recent one here lasted about an hour and a half. I had the depression,tired thoughts that take your mind on a roller coaster ride. I feel as though it's trying to zap my confidence from me. My fortune reads, "Don't let unexpected situations throw you."

There are a lot of people in the IBS forum that have it worse than me. Some have had IBS-diarrhea, anxiety, depression for 15 years non-stop. Some can't leave their houses or go anywhere because it's so bad. One girl had to go to the bathroom in a bucket in her house because her boyfriend was in the bathroom a bit too long.

I have found God which is what I needed to help me through this chaos. Today I've been barely have to get food down the hatchet because of the dry mouth. I have to drink water to flush it down. And when I am able to get food down it doesn't stay in my system long. I just end up with diarrhea right away. I believe if I continue to tough it out though and face things head on I will desensitize myself from the symptoms.

10/27/08, 9:20 a.m., at work

Morning has been rough. Was wide awake last night and woke up way before my alarm and couldn't fall back asleep because my stomach was ill again. My nerves are still keyed up but I am improving because I don't have that crying feeling that I had on my birthday. I had IBS-Diarrhea this morning for me. Upon waking up this morning I felt depression and a weird sadness. Worried I wouldn't be able to sleep again. Made me nervous thinking I'm never going to be able to sleep in again and will always have sad thoughts. Considering medication still since my mind is having trouble stabilizing. Still tired but if I was to go home right now I'd just roll around in bed and not sleep.

Still not calm like I want to be. Noticed when I lift weights I feel better for a while. Yes, improvement from a few days ago. No more dry mouth, able to eat some foods, no crying bouts like back then.

I think back to when I lived at my parents place for a year and remembered how the anxiety just suddenly passed. Main thing is I don't start in again with chew. I refuse to have any slipups now after dealing with this crap. I should have learned my lesson the first time when I went through extreme anxiety. The reason for me not taking meds is because I feel they only delay getting healthy. You'll eventually find you need to wean yourself off meds one day too which I read, can't be pleasant.

10/28/08

Slept bad again last night because of stomach acting up. Woke in the middle of the night to sprinklers going off outside. Today I've had 3 crying fits that come out of nowhere. I believe quitting the chew caused me to have Irritable Bowel Syndrome .When my IBS acts up I lose control and that's when I get that hopeless feeling. It's not like my sleep was horrible last night either. Went to bed at 10:45 p.m., woke to that sprinkler outside at 3 a.m. or so, went back to bed (even though my back felt hot) and slept until 6 a.m. and that's when my stomach got bad. I had gas & nausea and then diarrhea. I've had diarrhea all morning long again. It went away when I drank ginger ale. That's what I've been having the entire afternoon as it calms my stomach momentarily. When I think about myself not participating in functions I get sad and very down on myself. I wish I was able to eat and sleep normal. Blood Pressure = 113/79 Heart Rate = 94

10/29/08, 5:00 p.m.

I went with my wife to the grocery store. When we walked around the grocery store I began to get my hunger back. I actually "wanted" to eat and some things there looked tasty. Before we returned home we got a burger at Burger King. I ate fast so that I didn't lose my window of opportunity to consume something solid. It felt good going down and settled nicely. I felt relieved to know I had this and watermelon in my belly. I will eat more soon but will keep it light. I was sick today all the way from 6:00 a.m. to 3:00 p.m. which also ignites the depression because I feel discouraged when I am stuck being sick.

To wrap things up, was probably my most difficult day yet since quitting chew. I have been sick to my stomach the majority of the day. Yesterday I ate 1 turkey on sourdough with provolone cheese on the beach then at home I had baked beans. Something didn't agree and the day before that something wasn't agreeing as well. Tonight I consumed watermelon, Whopper, watermelon again, smoked salmon with crackers, chamomile tea, and a half turkey sandwich as well as a shot of milk.

This should promote healthy sleep tonight. Turkey provides that amino that helps induce sleep, tea helps sleep, and so does calcium in milk. I truly believe I have IBS and will do research this week. I've been reading stuff in the past about it and a lot of people complain of gas, diarrhea, nausea and other problems and as a result of that, lack of sleep so it just makes it worse on them.

October 30, 2008

Today was a good day for a Monday. Usually they are much more difficult but my mind stayed occupied. Today's food = 2 turkey sandwiches, applesauce, 2 hardboiled eggs, 1 orange, watermelon, water, then when I got home I had 1 chicken breast with peanut butter sauce followed by some more watermelon. Last night was the first night in a while I was able to sleep the entire night without getting up to use the restroom! I woke up at 6 a.m. and lay in bed until the alarm went off. Today was a better day for me though, I should be complaining about a thing. Haven't thought about chew until today. Just mild thoughts of what it would be like to have a big chew again. Would I get buzzed?

October 31, 2008

Today was another great day for me. No anxiety or depression. Felt good and felt like when I got to my lunch hour that I could workout strong. It feels good to have my mind back again. My stomach had gas today quite a bit and I didn't sleep perfect last night so I can just imagine how I'll feel once I'm done working out.

November 1, 2008, 7:50 p.m.

I woke up at 6:40 a.m. with bathroom problems. I know I'd have that because I ate pasta & lentils for dinner as well as Reese's peanut butter cups. I hate moments of depression or sadness today and been tired. I want to sleep in but couldn't last night because the rain noise, bathroom issues I had, and constant gas. Still not out of the woods though. Now I'm dealing with heartburn which I read in a tobacco forum that that is another withdrawal symptom. I have been craving sugar like crazy. It's part of nicotine withdrawal.

November 3, 2008, 9:00 p.m.

Today has been a good day for the most part. Around 3:30 p.m. I had a jumbo burrito which stuffed me up pretty heavy. I had a tired feeling for a few hours. I have been working out at the gym during lunch hour again. It feels good, but forgot how much food needs to be consumed. Already my calorie intake is like how Michael Phelps' is. I need to be careful not to overdue it. I can already notice my face getting puffy from all the excess food. I haven't had depression or anxiety but have had morning sickness recently.

November 8, 2008, Fourth month off chew

Today made 4 months off chewing tobacco! Did my blood pressure at Long's drug store today and its 101/72 and a heart rate of 62, which tells me I am starting to calm down finally. I had slept good last night and woke up around 8:00 a.m. this morning and was still tired but couldn't go back to sleep. I still have a little morning sickness, the gas has subsided quite a bit which helps me sleep.

Today I felt mild depression around 1:20 p.m. to 2 p.m. but took a nap for an hour and it went away. Today's food = bacon & eggs, 2 ice popsicles with cream in the middle, deep fried shrimp, 4 reeses peanut butter cups. I should eat healthier though.

November 9, 2008, 7:30 p.m.

I had a decent day today. We went to church, felt good, and had an okay night's sleep last night. Woke up at 8:00 a.m. and felt like I slept a full night even though I got up 2 times for the restroom to urinate.

Wife and I went to the car dealer just to "look at the new cars and wound up almost buying one. I felt some anxiety during negotiations because the dealer wouldn't let us walk away. He kept talking and talking and asking me if I liked this price or that price. I was beginning to feel obligated to buy a car! The anxiety stuck around until we got home and after all that, my stomach got indigestion.

November 10, 2008

Good day today at work. Had some minor anxiety with sushi for dinner at a restaurant we frequent but settled down after 15 minutes or so. I did my best to just roll with it even though I didn't like the discomfort.

November 11, 2008, 9:15 a.m.

Woke up with a stomach ache this morning and it serves me right because I ate almost an entire bag of Reese's peanut butter cups last night!

Chocolate contains traces of caffeine and my doctor. said to not have any caffeine during my nicotine quit as it can cause anxiety. So it's no wonder why I feel tired now. If I would have had a pleasant nights sleep, my body would have been sending the message that I can eat tons of sugar and be fine. I can't get over how much I'm craving sugar and chocolates lately.

Wednesday, 11/12/08, 10:00 p.m.

Good day today. I had some bloating problems at 8:30 p.m. after eating dinner (meatloaf & potatoes). But, I've been having gas come out so that's helping my stomach. I realized today that I no longer have that tickle in my ears. It's been about 3 weeks since I last remembered it being an issue. So this morning I hit the snooze button 2 times which is good because it means I'm sleeping more. No more racing heart in the middle of the night either!

11/13/08, 9:30 p.m.

I didn't sleep well last night. The neighbor's dog kept barking so when I woke up this morning I had to take it slow. Right now I'm tired and feel that I can just sleep as soon as I lay in bed. I had stomach aches through out the day. Gas is still a factor.

11/14/08

Got that extreme fatigue feeling at around 3 p.m. today. Felt tired and wanted to go home early but stuck around. When I woke in the morning I had a bit of a stomach ache. Went to the bathroom and felt better, but was still tired. I get tired fast lately. One minute I'm fine, the next, I need to lay down NOW!

11/15/08, 8:50 p.m.

Woke up with a stomach ache and had to constantly urinate. (This is a sign of being nervous about something.) Happened all morning so I couldn't go back to sleep. Kept waking in the middle of the night and my body was hot. Today was challenging but not overwhelming like a few weeks ago. I became tired throughout the day. I will be totally fine one minute and then the next, tired and want to sleep.

11/18/08, 7:30 p.m., 19 weeks off chew

Today was a good day. It was difficult waking up but once I was awake, things were fine. I NO stomach ache to talk about today. No anxiety or depression. I remember wondering why it wasn't coming around. That is good news. Actually was interested in vacationing again which is nice for a change and I bet my wife looks forward to hearing that. For dinner tonight I ate 5 slices of pizza. I am very stuffed now with some gas. I ate at 7:15 p.m. For some reason, ever since I quit chew, I can't say when to stop eating. I tend to over eat. When I used to chew, I'd always stop and have a chew or have less of an appetite.

11/23/08, 10 p.m.

We got back from Cayucos, California. We spent the night there. After we arrived last night though I got that extreme tired feeling that lasted a good hour. I suddenly got sick, felt like I wanted to cry and felt like I was going to let my wife down for coming all this way and being stuck inside the hotel. Prior to me getting this way, we had stopped off at a Burger King and had a whopper & fries then later we ordered two deep fried fish tacos. So I was pretty full. We had made a reservation for a nice restaurant that I wasn't sure that I'd feel comfortable in. We ended up taking her to an Italian place so she could have some pasta while I sat and drank water. It wasn't until we arrived back at the motel (approx.9 p.m.) that I started feeling good again.

We had originally arrived in town at 4 p.m. we ate at 4:30 p.m. After that is when I wasn't feeling great. I laid down for a couple hours, took shower. I know that we did do a wine tasting along the way on our drive which probably didn't help my cause. I learned that that sad feeling that I got in the motel does pass though, even though when you are in it, it seems like an eternity. This one lasted about 3 hours so I have to prepare for that when we go to a BIG vacation.

11/26/08, Wednesday, 5 p.m.

Tomorrow is Thanksgiving Day. I had some anxiety thoughts just now so I decided to write about it. Kind of like a fear of the unknown feeling. Frantic thought then faded away after a little while. Not as much sad thoughts as I was getting around my birthday. For today's food — peanuts, cheese sandwich, 2 oranges, 2 mini cheese cakes, peach tea, and 2 waters. Besides the oranges, everything I ate was bad for me and I know that when I eat clean, I feel better. But my problem is that when I do get hungry, I crave foods that are high in sugar and salt content now. I read that that is part of the withdrawal process. This morning when I woke up I was still tired so I'm hoping for a full nights rest tonight. It's no fun feeling like this before getting together with my family.

Thanksgiving Day, 11/27/08, Noon

Feel off today. Didn't sleep much last night because we ate turkey and stuffing at close to 9 p.m. last night. I woke up at like 4 a.m. with stomach problems that kept me awake up until around 7 a.m. Then after that I'd sleep in 30 minute intervals. Things still feel tough for me as far as being around people or making plans goes. Right now I feel run down, a bit sad, and worried that people will be concerned about me. I also feel I won't be able to eat because my stomach isn't right and I don't have an appetite.

Thanksgiving evening, 8:30 p.m.

Today was a difficult one for me. When we arrived at my families place my stomach wasn't right as people continued to arrive. I'd just keep getting sicker. Then finally when everyone started leaving, I was able to start eating a little bit. Two stuffed mushrooms, greenbeans, and some cherry tomatoes. When we arrived back home we did a two mile walk. My stomach felt fine during the walk but when we arrived back home and sat down, I felt sick again. I started thinking some very negative thoughts that just wouldn't get out of my head. Sad, negative thoughts that only produce a low self esteem.This thought has caused me great pain and anguish. I've had gas and stomach aches all day. I am tired right now and want to make sure I sleep the whole night. (Notice how nicotine withdrawal can even put a damper on a person's relationship.)

11/28/08, 10:30 a.m.

I'm not sure what's going on with me but I've been crying my eyes out and praying to God constantly. IBS is no joke and the past 5 hours have been hell. I don't know any other way to fight it other then praying to God. Pills just cause other problems and you get hooked on them. I forced down oatmeal today which was good but still not great. I feel once I get a full nights rest, I will feel better. I know that praying to God though does help. He has rescued me in the past with a simple ring of a phone. I would like another symbol from him now though so that I know he's listening to my cries. I feel better when I thing about the Lord. Today, I will listen for him & look for him. They say he's everywhere and I'm very fortunate to have my wife in my life who has taught me to pray. She is so sure He's up there that it brings me comfort and I trust that she's not fibbing. Heart rate = 127/69 Blood Pressure = 63.

Noon

Just got back from a bike ride, my stomach is still bad. All I've had to eat today was that oatmeal and a bit of applesauce. Just about 10 minutes after my bicycle ride I threw up. It came out of nowhere. I had so many symptoms today, nausea, loose stools, gas, noisy stomach, crying for no reason, emotional and now a little vomiting. Chewing tobacco must have really poisoned and masked a lot of stuff inside of me.

11/28/08, 1:45 p.m.

I'm doing a lot better and I owe it to God for answering my prayers. Yes, I still feel a little off and tired but I'm on my way towards recovery. I was able to eat Oatmeal at 10 a.m., applesauce at 11 a.m. and saltine crackers with peanut butter at 1 p.m. Then later I managed to put down potato & celery stew along with a few bites of smoked salmon. I can feel my mind going back to normal and away from the crying feeling I had been having through out the day.

It feels good having food in my system. It takes away that pain and scratchy feeling. My mind is back to not worrying about things again. Sure, that can change at anytime, but for now, I'll take this feeling. I noticed that when I have stomach aches, I fall apart and tend to panic, freak out and get sad. I start thinking it will be that way the whole time on vacation which is a deranged thought. When my wife and I went to Maui last year, I was sick on the plane flight over and then that weird depression when we arrived at the hotel. I was a big chewer at the time. So that just goes to show that chew can also make you sick too.

The second trip to Maui I didn't have a problem. The second trip was fine for me because I kind of knew what to expect and had my chew. At work I'd practice staying in my seat until 11 a.m., then getting up to use the bathroom. I also brought my sounds to listen to. On my trip with the wife to New Hampshire (across the Country) I was intoxicated,punch drunk which made the plane flights exciting. Problem is it made me sick the next day which really threw a turd in the punchbowl since I was going to have to be around her family. Dr. Gorman said that I may have a temporary intolerance to alcohol since my body is adjusting to life without nicotine. My body could be rejecting just about any substance let alone nicotine.

11/28/08

I am at a Church lawn right now. My mind is all over the place today. I only slept maybe 5 hours last night if that. You would think I'd be okay with that amount but no. I feel rotten actually. The thought of being around people today bothers me for some reason. I haven't prayed this much in a single day then this day. I have been crying for no reason what so ever. I wish I knew exactly what was causing this. Yes my chemicals are unbalanced in my brain due to quitting chew and feel like I just quit a day ago or something. I have been through a wave like this before. This time is more of a sadness, insomnia, and stomach problem whereas before it was more anxiety episodes.

Yes I need help but in the form of God, love, and peace. I need people that support me, not get unhappy with me. People that are willing to accept who this new person is and say that what I'm going through is all part of the process. I don't need those that look at me like I'm nuts or have medical issues. I don't need people in my life that will be upset with me. I no longer want chaos in my life. I would love to eliminate all difficulty from my world but that just would be crazy. I never remembered having all these problems growing up though. I do remember being over my friend Rich's house when I was 17, having drinks from his parents cabinet and I'd wake up sick the next day. That was normal though. As far as all this other drama though, I never had it before. I've had some minor thoughts of chew through out the day.

10:15 p.m.

Felt better at around 5 p.m. Ate turkey, stuffing, mashed potatoes and veggies at 7 p.m. followed by sleepy time tea at 9 p.m. I think it's safe to say that I'm feeling good. I owe it to God for answering my prayers. Had I not called on him for help I would still be feeling horrible.

11/29/08

I'm nervous all morning for no reason. I know it's from quitting chew even this far out. I've been trying to talk myself out of going on vacation. Stomach has been on and off all morning. Slept from 11:30 p.m. to 5:45 a.m. I have to remember that all these sensations are caused by my head. My wanting to cry **for** no reason makes no sense. It's just a chemical imbalance and my brain will eventually balance itself out. I know it is caused from quitting chew and it threw off my chemicals…

Everytime in my life that I've experienced this sick stomach feeling with nausea, I have suddenly got better. So what makes me think this time will be any different? I have also had bouts of depression, sadness and severe anxiety where my thoughts weren't right. I have also just snapped out of it from time to time. All this crap is in my mind and it's playing games with me. The anticipation will always be worse then the actual event.

Navy Seals go through "hell week" where basically they have to do all sorts of training and working out on 1 to 2 hours sleep and a power bar for a meal. I remembered one guy on the show sayind "I have no saliva or spit to swallow my food."

11/29/08, 5 p.m.

I just had my first anxiety attack in a long time. It lasted about 15 minutes. I still feel the jitters of it. I don't want to go on a vacation with my wife anymore but dread that I'll have to tell her to cancel. The way I've been feeling isn't good and being far from home would only make me feel worse. The hardest part about all this is letting her down. That's the last person in the world I want to upset. I love her so much. I dislike this feeling though of always having to back out of vacations. These days, I just want to be at the comfort of my own home. (Is withdrawal at its finest?Is it getting the person to lose control and change ones thought patterns of something that should be relaxing to something that would be a chore?)

It seems as though I have hit rock bottom. I have tried everything that comes to mind including praying to God non-stop and nothing seems to be helping me. My stomach is still not good and I can't put any food down. I still have moments when I feel like vomiting or crying for no reason. I'm scared to take medications because the ones I've tried make me extremely sad. This can't still be from getting off the chew now could it?

Yes actually it could from what 2 Doctors I have spoken to. Smokeless tobacco withdrawal can be equally or just as hard as quitting cocaine. It can also take 6 months to 1 year to be completely finished with withdrawals! Be Patient!

11/30/08, 10:45 a.m.

I've been getting bad anxiety attacks the last couple days. I averaged 4 hours of sleep the last few nights because of the upset stomach. See I wouldn't have these problems if I had chew to use.

This is really horrible.

Still have to force feed myself. Smells of foods being cooked around me makes me sick. I still have frantic thoughts still have fatigue. I'm almost certain though that if I keep things up and cancel on her again, we'll probably break up (notice how sensitive I became at the four and a half month mark of going cold turkey). I don't like hurting her as I stated earlier and this has been one of my worst days since I've met her. If I was a normal person again right now, we'd still be going to our trip in Belize. But right now, I'm not. I'm going through something strange that I went through a few years ago when I tried to quit chew. My heart rate = 131/76. Blood Pressure = 76. I'm considering taking antidepressants again. Thanks for causing me all the anguish chewing tobacco. You've been a real pal.

12/2/08, 10 p.m., Tuesday

So the anxiety and stomach problems went away at around 2:30 p.m. today. Five full days that IBS and drama hung around in this episode. It wasn't fun at all. I believe my turning point was staying on schedule and still going to the gym at my lunch hour. I was able to consume regular good clean food tonight that's solid. It's just what I needed because for the last 5 nights my stomach has been waking me at 4 a.m. and not letting me go back to sleep. I owe all this to God and keeping me focused. This felt like hell since last Thursday. Always having to force feed yourself and then the diarrhea just isn't right.

12/4/08

One week following Thanksgiving. No diarrhea to report, minor anxiety today. I can say I feel much better now then I did on Thanksgiving Day. Today was difficult at work even though I slept well last night. I woke up this morning with bloodshot eyes and this weird depression. Not Horrible cry potential depression, but the kind that makes you take inventory on yourself, and causes you to think about how you are doing more often. I have meds ready to take but I won't take it if I am feeling like I am now which is better, stable thoughts and ready for bedtime. I don't think I'm yet willing to take the punishments meds have to offer me in the beginning. I know that my mind is still in the healing process though from quitting the chew, it's just got to be. I lived at my parents for a whole year there (well 10 months) and was done chewing at least 2 months at the old apartment before moving in with them. That means I quit chew for a whole year but then started in with nicotine again the moment I moved out. I do remember going for a quite a few months there where everything was in control and I was feeling wonderful again. I remembered being able to handle stressful situations that I would have otherwise not been able to fresh into my quit.

The depression, anxiety, and insomnia vanished though and remembered that my difficult times were coming home there. I didn't like the idea but would go out to the chicken coop and feed the chickens to get my mind off it all. My problem's faded fast though and was able to relax as time went on.

12/5/08

I just got back from 3 mile walk. Today I've had a good day. Got to work on time, was able to eat a little granola bar, salad, 1 slice pizza,two oranges, 1 can oysters, 1 chicken leg, 5 chicken wings with blue cheese. No crazy anxiety to talk about. I'm not getting along with my wife much so we'll see how this gets my stomach. She doesn't realize that when she gets mad and stops talking to me that I get sick over it. When I get sick, she gets more frustrated, making me even worse! I really hope I get better! The Dow is at 8635. It's been dropping so rapidly. Even my money is going south now. I'm not getting anywhere it seems.

12/6/08

Today I was able to fall back asleep after waking up at 4 a.m. to eliminate the hot wings that I ate yesterday for dinner. I fell asleep until 8 a.m. then fell back asleep for a couple more hours. I still didn't feel 100 percent today. I had those frantic thoughts that would last just a few seconds. I tend to think that I won't be able to sleep in the coming days and that I will have another mental breakdown. I still get worried when I go for 5 hours and still feel full from a regular sized meal. I still believe I am withdrawaling from chew. Six months is actually when you start feeling good again, or so they tell me. If I am still messed up with anxiety and depression after that time then I know I can tell them otherwise.

I know that when I quit chew last time all this same stuff happened to me too. It's not a surprise it's happening again. It would only seem logical that the reason for insomnia, depression, sick to my stomach, IBS, and anxiety is from quitting chew. I did the tanning booth today and that made me feel good about myself. I read something that it helps curb depression. I want to see if it's true or not. I have noticed a change in my attitude this evening. We'll see how it goes long term. Food today — half a cantelope, burger and fries, 1 bowl cereal, 1 chicken sandwich, and tea.

12/07/08 9 p.m. Sunday

Today was a challenging day because I was tired when I woke up. I can feel anxiety surges in my chest come from nowhere. Those thoughts I was getting were negative. I'd think about my Grandpa and what he'd think of me if the next time I came over and wasn't able to eat like I wasn't able to on Thanksgiving. Woke up this morning though and didn't have a horrible stomach ache but did have to constantly urinate which is a sign of being nervous. I went to church, then to a restaurant with the wife for breakfast, then for a short walk. During the walk I got tired and sad which is unusual for my walks. Came home and laid down for 30 minutes, felt better. Went to the movies and felt good even though I had moments where my stomach would feel upset. My stomach was off even when we did our walk this morning. Today I ate half a cantelope, fruit pieces, 2 sips of V8, 2 eggs poached on avocado, tomato with a hollandaise sauce on bread. Then for evening dinner I had a chicken sandwich, watermelon, and a bowl of cereal. I keep praying that this anxiety depression will vanish or go away. It has been improving though since 5 days ago. Each day is slightly better, and stronger. I am learning how to cope with my feelings better. I remember that at church today I had brief moments when I felt like I could cry but not nearly as intense as the last Sunday I was there when I had severe strong feelings that I was going to cry then ended up leaving early and getting out to my car and crying my head off. It made no sense at all to me why I was crying.

Its all part of nicotine withdrawal!

12/09/08 5 month anniversary of no chew!

Today made 5 months no chewing tobacco. Man what a struggle it has been. But I'm not going to lose my quit now. I hear a lot of chatter around me of people saying they are unhappy with me over things because of all the anxiety the quit has caused me. Today had its ups and downs. Went to work and after the gym I became extremely fatigued and started in with those negative thoughts that became depression for 1 hour. When it comes along, it isn't as strong as it was a month ago or the bout I had a week ago. Plus the crying fits that I was getting disappeared. I have been able to eat for a change with only a couple hours in the day when I didn't have an appetite. During those moments I was thinking negatively which was what caused my biology to get like this. They say that you don't really start feeling good until 6 months quit. I remembered Dr. Thompson saying it can take a couple years before your body's chemistry resets itself when getting off nicotine. I'm glad I didn't take antidepressants. That would have been rough and would have changed my whole life. I'm glad I listened to myself and not other people regarding this subject.

12/10/08, 8 a.m.

I've been a little sick to my stomach this morning. I am nervous for some reason too where my teeth chatter. Plus I had diarrhea this morning. Last night I ate chicken rotisserie from Safeway & stuffing with Non-dairy butter. Last time I had stuffing was before Thanksgiving and I'd gotten sick as well. I had a lot of gas this morning followed by the diarrhea. Had interesting dreams last night and I woke up a couple times; wide awake. Can't stand when that happens but can't stand it even more when I can't fall back asleep and just lay there looking up at the ceiling. Glad I'm able to rest more today. When I was a chewer I had no major problems with the sleep.

So far I was able to put down a peach cup and having a cough drop currently. I have cough drops because it keeps me from having dry mouth. This has been a rough morning for me. I do feel more rested then the last few days and am thankful for that. I know now that I just have to keep fighting and if I'm not going to be willing to take meds then I'm going to have to deal with it head on like Jeremy said in class.

Just like the other night in the group meeting when they went around the room and each person took time to speak. I went and spoke after the person before me took forever and started crying. She really took a while but I think it made her feel better.

My anxiety level was getting higher by the second I recall but then once I said my part, I felt great! Like Mr.Kernion said, "If you aren't feeling good, and are tired, sick to stomach, can't eat, nervous, still go through with it and stay the course because you feel great you did whatever it was afterwards."

9:30 a.m.

Still not feeling so hot and am having a tired spell right now. I can't wait for this to pass. Hard to say which came first — the stomach problems, or the depression/anxiety. If I understood all this more I wouldn't be writing all this stuff. Yesterday made 5 months no chew. I'm glad I didn't start in with that again. I refuse to give up my quit.

Last night my wife and I went to bed at around 9:50 p.m. I feel to sleep right away. I must have been very tired! I plan to continue to go to bed around 10 p.m. every night until I start getting successful naps.

Just worry about whatever it is you are doing and avoid the rest of your thoughts. On days you feel alright you just think about the moment and not future events. On days I didn't sleep well and am depressed, I've found I think about how I might possibly let loved ones down; like I've let my wife down 4 times now. I have one more month to go before I can officially not blame anxiety and depression on nicotine withdrawal anymore.

This feels like its lasted forever. Everyone kept fighting for me to quit so I really don't want to hear it from them. I had no idea that when I quit chew this time that it was going to last this long. Five months out and I still have the loss of appetite, diarrhea, insomnia, gas, fatigue, tickle in ears, depressed, lack of interest in hobbies. I would really like to figure out the meaning to all of this.

3:06 p.m.

The discomfort comes and goes, I will have intense discomfort in my stomach and get really tired, then it fades and I feel okay. Then just when I think its all gone and I'm out of the woods, it strikes again. The sick feeling will bring on frustration. Although this stuff feels like a nightmare right now, there is light at the end of the tunnel, I just have to have faith. Every other time that I've gotten sick like this, I have eventually snapped out of it. My brain will eventually get tired of worrying and I will get relaxed again. Right now, its easy to worry about something and get sick over it but pretty soon its going to get hard worrying about something and get sick because I'll be pooped out of excessive worrying. It's a proven fact through me!

Besides the worry wart stuff, I just want to feel normal and avoid meds if I can. I'm not taking any but have Celexa up on my shelf if need be.

I do believe I've grown a lot from what I've experienced in the past 5 months. I've learned that when you get sick to your stomach, it passes . When I get depressed, it fades. When I get anxiety attacks, they dwindle away. And when I get angry, I eventually become happy again. When I'm fatigued all day, I can still get through the day alright. Yes, it absolutely does suck but I really have no choice in the matter unless I want to start in with meds which I don't.

12/11/08, 8:00 a.m.

Got up this morning and had a bit of a stomach ache, went to the bathroom. Noticed I had a problem and was stressed. Hoping this is the end of all this drama now. In a couple days from now I'll be strong again. My stomach is already beginning to feel better just thinking about it. I'm going to have something to eat. I had a peach cup. I pray that I have more windows of opportunity to eat today and that my body allows me to consume solid foods. Had a moment in the shower this morning where I felt as though I could cry but I held it in. It just came out of the blue. It came at a moment when I was frustrated with the way I was still feeling. I thought that the feeling should have passed by now. Stay the course is what I just need to keep doing.

Tobacco did this to my brain and I'm going to allow this time for my mind to adjust.

9:10 a.m.

I'm still nervous about eating anything solid. Afraid it will make me sick or give me bathroom problems. Had a moment in the lunch room just now where my stomach started hurting really bad like I was going to vomit. I can't wait for the day that I'm calm, mello and easy to talk to for others. Remember, no one can tell you are nervous that has got to be a myth.

These waves come and go. When I hydrate myself I feel good but that feeling gets followed by anxiety because I still am not able to eat unless I force myself. I realize I haven't been myself for many many months now but that's just going to have to be the way it is. I need to stop being impatient with myself. Last night we went to bed at 10:15 p.m. but the neighbor stayed on the phone talking Indian until 11 p.m. so I couldn't sleep. Went to sleep around that time and slept until 5–6 a.m. So I had a good 6–7 hours rest last night. The gas in my stomach is what woke me up this morning though.

4:27 p.m.

So I have been feeling better ever since 1:45 p.m.. I took a nap at lunch today instead of working out and felt a huge difference in my attitude. At 1:45 I quickly drove to Safeway and bought some chips, sushi, and water. Ate the chips & sushi and felt great afterwards. It feels good having some real food in my stomach for once!

12/12/08, 8 a.m.

Woke up this morning and stomach was off again. Drank water then got up. My mood was so-so this morning. I didn't have the feeling of wanting to cry but was still a little down. I really need a day that I can sleep in. When I'm able to sleep in all the time is when I'll be able to say I'm getting better. What I've noticed lately as a trend is that I need to keep preparing my stomach to eat. I need to have some tea or a cough drop before eating anything. It's extremely frustrating to have a stomach problem for this long. Good news is that every day I'm like this is another day that I'm desensitized from this suffering.

Last night I went to sleep at 10 p.m., woke up once in the middle of the night to urinate. I went back to bed, woke up way before my alarm clock. My stomach was giving me problems and could tell I needed to use the toilet. I probably had around 6 hours of sleep last night. I'm over due for an 8 hour sleep. Depression is still hanging around and keeping my mind thinking. Gas still erupts in my stomach, lots of burping.

12/13/08, 9 p.m.

I made it through today with a lot of progress. I started the morning with gas, diarrhea and slight depression. Instead of me going to the gym at lunch I decided to close my eyes in my car. As a result of the nap, I woke up hungry with an appetite to eat something solid for once. So I made my way to a local grocery store and bought some sun chips & California rolls (sushi) and a small orange. That was my first regular meal in days.

12/14/08

I'm sick of feeling like this. Nausea is non-stop. Last night was a rough one so I have to remember not to do any late night eating. I'd rather just have a little bit of food in my system then a bunch like last night. Holiday's tend to really stress me out now. Everyone expects me to be somewhere with a certain happy attitude, giving the right priced gifts and not letting anyone down.

It feels like I have the flu though because I've felt sick for so long now but no, its all from my quit. I never acted this way before my quit. Life without nicotine is weird. I'll just pray that these feelings will be gone and I'll be able to sleep in and not have to wake in the middle of the night to go to the bathroom. I felt good doing our walk today but other then that, things just get to me easily now.

12/15/09

Depression is still the main problem. Tired, fatigue, negative thoughts are still there but I'm trying to take it all lightly. I was reading the quit nicotine group this morning and it looks like a lot of people that quit around my date are experiencing the same symptoms. Plus I don't do well during the holidays. There is a lot of pressure to be your best. I still feel that surge in my body every so often through out the day that just comes out of nowhere but if I look at the big picture, it's because I took tobacco away from my mind. Its weird too because I dread being at home. I'm never sure when I'm going to walk in the door and she will be disappointed in me because I'm not planning any trips to anywhere right now.

I also noticed that I have extreme gas in my pipes in my stomach. When I don't burp for a long period of time I get the nausea that makes me feel like I'm going to vomit. I blow up like a balloon. Lately I've been having depression episodes from lack of sleep. If I had a full night's sleep this depression thing wouldn't be around, I can pretty much guarantee it. Sleep is the Key….Forced myself no to nap at all or eat after 8 p.m. as stated in Dr. Maas's book "Power Sleep." I am determined to get a good rest tonight. I feel optimistic about tonight at the moment but I don't know when that will switch to problems and negative thinking again.

I've noticed that when I'm sitting down for long periods of time, gas will bind up in my stomach so when I get up and walk around, I can feel the gas moving through the intestines, processing once again. But when I get the stomach aches, they make me panic and sometimes I get anxiety attacks from it. There were moments when my brain would need a lot of convincing to assure it that nothing was wrong and that this was all good symptoms of my body healing from an aweful drug. Going for walks helps mentally.

Walking also helps with indigestion.I've noticed cough drops also help with stomach indigestion for a little while and keeps saliva in the mouth. I am not sure what is in a cough drop that soothes my stomach but it works. My stomach settled down tonight around 6:30 p.m.

12/16/08, 8:27 p.m.

Had a good day today with virtually no depression or anxiety. I was able to eat food too. I didn't sleep excellent last night either. It goes to show that brains can fix themselves on their own without meds. So just pray and let Him do the rest. The only time I had a little bit of depression & anxiety was when we went to dinner at Burger Pit. All those anxiety symptoms and not being able to breathe smoothly made things difficult. I was taking shallow breathes and then try to make up for it with one huge breath. Then when we left I became normal. And I was fine dealing with it in the restaurant as well. Two days in a row of feeling good. I've been on a roll. I haven't thought about chew, I no longer think about it.

12/17/08

I had another good day. We ate well, turkey sandwich, granola bar, sunflower seeds, orange, burrito, 2 waters. And when I had the burrito I became very sleepy. This feeling lasted for about an hour. I had to keep walking it off because I'd have fallen asleep at my desk. Mentally I'm feeling stronger today then a week ago. I am eating plenty, sleeping good, thinking clearly. We went to church tonight, no anxiety whatsoever. I didn't have any for the entire day which is a first for me in weeks. I can't remember the last time I was this comfortable and relaxed in my body. I have a feeling I'm getting better because I'm starting to have Positive days in my life again like this one!

12/18/08, 7 p.m.

Wife and I are in an argument over little stuff. They say not to sweat the small stuff, because it's all small stuff but it seems to really be on my mind. Took my blood pressure today and it was 122/73 heartrate = 53 so I'm normal. I know that any little thing can set me off these days. Christmas is only a few days away and I get nervous around this time because of trying to do too much or please everyone.

12/19/08, 9 p.m.

Today was rough for me. Left work at 2 p.m. because my stomach felt sick and had depression that lasted for a while when I got home. I started crying like crazy and felt horrible. I couldn't escape it either. This lasted until I got to my parents place (6 p.m.) I was looking at my old diary from when I quit a few years ago and noticed how the time frames for when I was going through a bunch were very similar to this quit. I should have learned my lesson from what I'd wrote down the first time. Chewing tobacco is no laughing matter! I am sure that if someone was to do a report on the amount of people that committed suicide that quit chew, they'd realize what a serious thing it really is.

12/20/08, Saturday

I woke up around 5 a.m. keyed up again like my engine was started. There is absolutely no reason for this. My stomach was in knots and had bathroom problems. Today I made the decision to start medication. Uncle Hugh takes the same stuff and says it takes about a week for it to work. I can't stand when I get like I'm going to cry suddenly and for no reason. No One Died! My wife took off to see her family on the east coast and is happy to go see them and everyone seems well. I cancelled 2 trips in the past 3 months with her because of this anxiety depression stuff. I've had enough of me doing this crap. I don't want to ruin my marriage either.

Like my friend Rich said, there are huge numbers of people on antidepressants these days. Rich said he is fine 7 days a week now from taking meds. I sure hope this stuff makes me happy again though. There's one thing I won't miss and that's these sudden crying fits that I keep getting.

My dad said taking this pill should be like taking a Vitamin C. Don't think into it very much and that this shaky feeling will go away in a week. I sure hope he's right. I have been a nervous wreck today. I went to visit my Dad at my Uncle's place and kept getting those feelings like I wanted to cry again so it was hard being around but my Dad insisted I come by regardless. I came home and took a two and a half hour nap afterwards and it was hard to even wake up. This med stuff is really something. One minute I'm feeling good and the next I'm thinking this stuff ain't for me. This morning I've taken half the pill as prescribed with food.After that I ate more at lunch time even though it made me ill. I can't wait to feel normal again.

This has probably been my toughest day yet though. I still am hoping that all this suffering is for a reason. There must be an end to this somewhere. This is nuts. Cast thy burden upon thy Lord. That is what I'm doing. I know He can help me. I will continue to pray and not give up hope. I have never been a praying man until now. I've decided its best to do this method instead of the medication method. I think I'm better off riding this out on my own. Wife should not get frustrated with me though and be patient. She isn't the one that had a nicotine addiction for 16+ years.

I've noticed my best times of the day are evenings after 7 p.m. lately. Between 7 p.m.and 10 p.m., I'm at my best. Not sure why but I feel calm and it feels okay to be tired and sleepy at these hours. Like right now I'm not feeling too shabby, it's cozy in here, cat is relaxing in front of the heater. I have decided yet on if I'm going to take the medication again tomorrow. If my mind is stable then I'm going to stop but if I'm going through another day like this one, I'll have no choice but to do it. It's really hard to say if meds are right for a person or not. There are so many side effects in them that you don't know if you are improving or having other problems. Plus, once you are on them, there's no turning back, you can't just get off them and are mostly habit forming. So that mean's you'll have another crutch to fall on now that nicotine is gone. I don't want to go that route.

12/21/08 2:25 p.m.

Today has been one of those days. Last night I went to bed at the regular 10 p.m. and fell asleep. I woke up at around 4 a.m. with stomach issues. I tossed and turned for about 3 hours, getting up to go to the bathroom a few times. My life feels a mess when I struggle in bed or get stuck on the toilet. I have been praying to God like its my last day on earth. This sickness I have right now has changed my life. I remember a time when I believed there were no consequences for our actions down here. Now I realize that there are. I continue to repent of my sins to God and will continue to. He has begun blessing me like stopping my indigestion and diarrhea, he's also allowed me to keep fighting against anxiety & depression.

Of course my mind sometimes forgets that its Him that is healing me. I start trying to fix things on my own or trying to do what people suggest. Cast thy burden upon thy Lord as I stated earlier. Never in my life have I had this tough of a time of it. I'm having a tough end of the year and am super overwhelmed. I'm hoping it's just because of jitters and Christmas. I remember last year I didn't want to be around anyone either because I wanted to be alone with my chew.

I've noticed a new trend though lately and that is that my toughest times of the day are in the morning when I don't have anything to do. As the day moves on I get a little better and by 6 p.m. I'm usually pretty strong. Strong enough to eat a little something Blood Pressure 122/86 Pulse = 99

I have eaten light and clean today so there should be no excuses to the foods consumed. I've been feeling better now as the depression came and went. It feels nice to have relief. It makes you appreciate the good moments. Thanks to God.

12/22/08

Today was tiring from the moment I woke up until 5 p.m. or so. Went to my Grandparents place before going to the social anxiety class I attend. Class was good tonight. Got to talk about a Connecticut trip I'd like to take to see my wife's family. I'd like to do this trip in 2009. Also I got to feel that anxiety in class. It was a feeling that the instructor Jeremy Kernion wants us to feel so they can desensitize themselves to exposure. Today I had to force food until around 2 or 3 p.m. Then i could eat easier as the day progressed. I had Activia Yogurt today which always soothes my stomach. It also has Probiotics in it which helps eliminate bad bacteria in a person's stomach. I'm sure I have a lot of that from all the years of chewing Kodiak!

For dinner I ate solid foods that would be good for the mind including almonds which are suppose to be beneficial.

12/24/08

I have entered a trend lately and I don't know what to think of it. I've noticed that I now get all my anxiety in the morning and it tapers off through out the day. This morning for example, I woke up with it. It was like half anxiety–half depression. My thoughts at the time were that I would not be capable of traveling with my wife to visit her family EVER. That thought spiraled in my head, did its damage in making me even more unconfident, after that it vanished!

I try to think positive about stuff as I have been able to go on trips with her in the past. I've been all over with her before I quit chew. Been to Maui, Bahamas, New Hampshire, LA, all over. That stuff isn't close by either. How did I manage my mind back then?

I start thinking I'm a chronic worrier though. If it's not one thing then it's another. Let's back up for a moment. When she and I first went on a date, I had anxiety but it was because I was trying to quit chew for the first time and about 3 weeks into it. It has now been 5 months since quitting chew. I don't remember being this much of a writer about this stuff (notice how I started doubting myself).

My point is that even though I have been backing out of trips or backing out of fancy restaurants, I still feel I've been a good husband to her. I have still able to do things that I wasn't capable of doing when I first quit chew. I like going to church now which I never used to do before I quit. I love going on hikes with my wife even though in 2008 there were many occasions that she didn't want to. I never got mad at her for it though. She hasn't ever forced me to go to parties or nothing. And even for a friends wedding that I was freaked out about, she backed out to calm me down when I was in my first month quitting chew. So if I was married to someone that was always upset with my actions then I would end up being sick all day everyday. I'm glad she understands my issues a little.

4:23 p.m., Christmas Eve

Today has been filled with depression, fatigue, and anxiety. I keep wondering when that day will come when this stuff won't be around anymore. I remembered Tony (also a chewer) from my gym saying it will come, then go away for a couple years, then it comes around again, then goes away. Another friend Billy said he used to get anxiety attacks "in his thirties" after his chewing days also but said they just stopped all of a sudden. Good thing is I can still do lots of stuff. I've learned that I can still operate and function during times when I have anxiety and depression. Lately I've been able to go shopping and wait in long lines, be social with family, get my haircut by a loud mouth hairdresser, talk to friend on the phone. Go visit my Dad in the hospital, go to church, go to movies, go to restaurants like

Mimi's, Bella Saratoga, Burger Pit. So I am able to function, oh and I was able to handle work. My problem has been having difficulty sleeping normal.

12/25/08, 9:40 a.m.

Woke up and didn't have that dull depression feeling in my head nor did I have the anxiety. God is hearing my prayers. Even though I'm still not perfect, I can function today. Still have slight sadness which I feel is different then depression. My thoughts are still not totally calm yet. Once I get a week or two of solid 8 hour sleep, I will be on my way again. When this stomach calms down it will allow me to sleep through the night and not need to wake at 6 a.m. to use the toilet. I'm very thankful that I don't have the jitters, trembling, or freak out feelings like I did last Saturday when I went to see my dad at Uncle Tom's place.

One thing I need to learn is how to handle my tired thoughts. If I'm not going to sleep well most of my nights then the next day I will have those challenging thoughts. That's when it's nice to be able to control those feelings. There is a way of doing it, I just haven't mastered it yet. I'm still thinking all this insomnia, depression, anxiety, stomach tension, jitters, crying and mentally tired physically restless feelings are all caused from getting off the chew. Even though it's been almost 6 months, the circuits in my brain are still attaching themselves. I remembered an old timer who had also quit nicotine tell me that we have to give our receptors a change to get adjusted to the old life style of operating without a substance.

I get in these moods where I get really tired and just want to be left alone. Hard to talk to people, even people close to me, like my wife, best friend, or Mom. So yes, this new way of living is continuing where my mornings are filled with depression and anxiety and it hangs around until about 4 p.m. or so then fades away for the night. Today at my parents I was very sensitive to loud noises and the TV was blasting. Everyone was very happy except me. I didn't even want to answer questions and just wanted to sit and read. Sometimes in the morning I just want to be that face in the crowd. He's the guy that no one really pays much attention to nor cares about. I'm confident though that once 4 p.m. rolls around I start to get better.

Funny though because the trend used to be that I was feeling great in the morning and then terrible on my drives home from work. Now it has flip flopped on me. Is it because there is nothing more I can do in this day? Do I feel guilty in the morning when I have nothing planned for the day?

The other trend change was also a few months ago when I'd get the depression after coming back from lunch break between 2 and 5 p.m. So I've had the trends change to all of the following times: 5 to 8 p.m.; 2 to 6 p.m.; 8 a.m.-11 a.m.; 8 a.m.-3 p.m. I've even had spells where it has woken me up at 5 a.m. and I can feel it in my head brewing. Can't go back to sleep and get stuck in depression that way too.

12/26/08

Went to work today and had mild depression from 8 a.m. to 10 a.m. and that was it for the rest of the day. Ate normal and had normal stool for the first time in about ten days. Settled down a lot too and had peace finally. I was even thinking about traveling again. Had a thought creep in to my brain today in the car about chew but that's as far as it went.

12/27/08

Felt good starting around noon today. My appetite came back around 2 p.m. Grandma made me tons of food. Felt normal for the whole afternoon. Even started wanting to travel and had a second day of good thoughts about it. Traveling in 2009 I'd like to go to New Hampshire, Tahoe, Bodega Bay, and Reno. Any other trips would be icing on the cake. It's good to have pleasant thoughts about doing stuff again instead of dreading good times.

12/28/08

Good day today with no anxiety or depression. Maybe a little boredom now but I can deal with that anytime. It was nice to have this full good feeling all day today. I can't wait to feel good about everything again from traveling to getting together with groups of people.

Lately I've been dying to travel but also been craving chew the last couple days. It's been a while since I've craved tobacco because I was so caught up in my anxiety and withdrawal symptoms that now that those sensations are gone, my mind seems to be looking for a fix of some kind. Only 11 days until I reach 6 months cold turkey.

12/29/08

Felt good all day. No depression or anxiety. A little tired after eating taco bell at lunch but I didn't have depression that follows like before. All signs are pointing to brighter days ahead. Also, I asked Dr.Gorman if depression would still be a product of quitting chew even 5 months out like I am and he said, "Yes" and that nicotine attacks a lot of receptors in the body. People were right though and that is that 6 months is the turning point to feeling good again. Praise God!

12/30/08

Day was good until around 3 p.m. At that time I got tired and worried thoughts entered in. I closed my eyes at my desk for a few minutes until depression passed. I have emptied my system quite frequently today. Not sure if it's from Taco Bell from yesterdays lunch or the 2 pieces of white chocolate I had today at 10 a.m. I know I had bad gas and indigestion all day today. I actually think it could be from the chocolate though. I heard there is more caffeine and sugar in white chocolate then there is in Milk or Dark Chocolate. (Note — Studies have shown that after quitting nicotine, certain foods become difficult to digest. One of the most difficult foods I found was chocolate.)

12/31/08

Today was challenging. Was reading what some people were saying in a quit nicotine group. This one guy, Bryce, is 9 days ahead of me in his quit and he has that inner "peace" that people claim begins at 6 months. Eight more days to go until I get there too. That is a huge goal for me. They call it the half way house. When I reach a year then things will really be better. I had difficulty at 2 p.m. today that lasted on and off until 3:15 p.m. That tired feeling came back around 6:15 p.m. to 7:30 p.m. after we'd eaten dinner then faded.

1/1/09

I didn't sleep that great last night. I had bathroom problems beginning at 6 a.m. We ate clams and virgin margaritas last night. I felt some depression when we finished eating. I could have eaten too fast, and clams could have been too rich for my stomach. I don't really want to eat which scares me a little bit because I don't want that to be the start of a whole new cycle. Food smells don't seem appealing at all at the moment. I have also noticed that I've become a very quiet person in the morning. I become more social as the day moves on. Usually on weekends I've noticed I start to become more interested in things around 5 p.m. or so. I want to make it to 6 months cold turkey and see what that actually feels like. I only have seven more days. I am so burnt out on feeling sick, insomnia, depression and anxiety. This morning I woke up to a bit of anxiety. I started getting worried because my wife bouth tons of clams & oysters last night. We didn't eat all of it last night and so my mind thought "all of those leftovers are still in the refrigerator will get spoiled" and she will be let down. Now that I'm awake and was able to eat bacon and eggs, I'm not as worried. I'm still praying for a solid nights rest. I know its going to be around the corner. There has to be a way to eliminate worry from my head. Christmas is past and so is New Years. All gifts have been distributed successfully.

1/1/09, 4:30 p.m.

Back from the beach with wife. Was going through spells of feeling extremely tired followed by not wanting to consume any food. Then I'd get hungry when we'd walk at the beach. My mind is playing tricks on me. When we got home at 4 p.m., I took a nap from 4:50–5:30 p.m.. I didn't sleep but I did close my eyes for a bit. It felt good to settle down without the TV going. My wife just told me that it isn't so bad living with me like this and she told me not to worry about her so much.

I do what I can each day that comes and today was a curve ball thrown my way. I am not used to feeling depressed at 4 p.m. The norm lately had been depression in the morning and feeling good from 5 p.m. on. Today that has not been the case. I have to remember what I read about chew and that is, a 2 can a week habit is equal to two packs per day. As far as the amount of nicotine goes, I had way over the regular chew every time. Honestly, I would go through 6 cans in a week like most chewers. And I did Kodiak which has the highest levels of nicotine then any of the others, even Copenhagen.

I am proud that I quit chew, I just want to know that I'll have a better quality life because of it. Like what I mean is, there has been some difficult times in this quit that I've thought would have been a better idea just to start back in with the chew again. I was having such chaos and turbulence in my head that I felt like suicide was just around the corner for me. After watching the movie "The Bridge" of people taking their own lives and actual footage of this, I didn't rule that idea out in all honesty.

Out of a scale of 1 to 10, 1 being weak and 10 being severe, I would rate today's depression & anxiety level a 6 whereas it was an 8 right after Thanksgiving Day. That was a hard time to function for me. It was so hard to do anything at all.

1/2/09, 2:15 p.m.

We just had lunch with my Mom and really enjoyed ourselves. I get happy knowing I made my Mom feel good because my step Dad doesn't really spoil her ever.

This morning I woke up to an upset stomach that woke me up at around 6am this morning. It was that same bubbling stomach where I have to press down on certain parts of my belly for the gas to pass through certain parts of my intestines. I woke up and had some depression and felt as though I could cry. My mouth was dry too. I brushed my teeth which helped my stomach a little bit. I went and had a blood test today at 10:30 a.m. I was nervous as usual but I didn't mind because I knew I'd be getting something accomplished by doing a blood test.

I was testing for some kinds of stomach diseases that are hard to pronounce. Pretty sure I don't have a problem and that all this stomach stuff is from my nerves and part of the quitting nicotine process. Eventually though, if I keep getting blood tests, I'll not worry about that so much. Like this time was less difficult then the one 2 months ago when I had my physical and I felt good about just doing it instead of postponing. As soon as I woke up I told myself, "I'm going to do this!" Afterwards I did a blood pressure test which came out 125/79 with a pulse of 75

I'd like to get that checked as often as possible. 125 is considered pre-hypertension but everyone says that that's normal. The rest registered normal. I think it has to do with me walking through Long's drugstore just before. I was able to have Activia Yogurt, salad, fries and a steerburger from The Burger Pit.

1/3/09, Saturday, 6:15 p.m.

I slept in last night until 11 a.m. this morning. I originally woke at 8 a.m., used the bathroom, brushed my teeth, fell back asleep 2 times.I had 2 horrible dreams about a couple fearful situations.

One was a dream about me having to walk past a pitbull that was chained up and then another was a dream of having to walk by a large number of pitbulls chained up. I understood the message my subconscious was sending to me. One was a small hurdle to get past, another was a big hurdle to get past. In the past I had been attacked by a pitbull so they've always been frightening to me. It's a huge fear that I should attempt to get over. I know that not all pitbulls were the way this one was.

Basically what it boiled down to is I have thought about desensitizing myself as far as traveling goes by going to Reno, Nevada with my wife (it was like walking by a pitbull chained up.) I've also thought about going with her to Connecticut and being with her family (that would be a huge obstacle to overcome like walking past a number of pitbulls.) Even though her family are some of the nicest, entertaining people I've encountered, this is the way my brain is thinking about things these days. What has chew done to me and my thoughts?

I was in a decent mood today, we went and saw my Dad at the hospital recovering from back surgery, then grocery shopping, and to Stanford Museum. After casually walking through the museum we got back in the car. As we drove off I got extremely fatigued. It was 3:15 p.m. We got back and I took a nap from 4 to 6 p.m. I woke up to some sadness feelings. Still tired afterwards and didn't have that nice appetite that I look forward to. Food eaten today—activia yogurt, 3 strips of bacon, 4 sausage, 2 eggs, water, red grapes in the evening. I was very happy to be able to sleep in for once though. Not sure if I should have taken the nap this afternoon. Hopefully I will still be interested in falling asleep tonight. I keep trying to remind myself, each day that passes I get farther and farther from chew. The more distance I put myself between chew, the better I feel. Today felt less difficult then yesterday, yesterday felt less difficult then the day before. BUT, if everyone in a while I'll have an extremely difficult day that may linger for a few days and I can't get discouraged by that.

Today's blood pressure results—117/72 Pulse—67. Waited 5 minutes then took it again and was 109/70 Pulse—66

1/04/09

I know today may be a difficult day because I slept light last night. It was because I took that nap so late last night. I was up tossing and turning for 3 hours. So when I did sleep it wasn't heavy. Right now my eyes burn a little bit and am a bit groggy. Good news is I didn't get woken up by bathroom problems this morning nor did I have a massive upset stomach. I did have gas still but it didn't make me sick this time as it was passing through my stomach. Like Uncle Hugh said about the crying fits, you kinda just have to ride through some of these emotions. Or like on celebrity rehab where that girl said she felt like she was riding through one emotion that lead into the next. Its kind of how I feel lately. I am truly blessed to get where I am today though. I have had so much weight on my shoulders lifted off and it's nice to know that these thoughts, feelings don't just stay the entire time. I get sad then normal, I get tired, then awake… I get sick, then I feel better, I stay up all night, then I sleep for a time. My biggest concern was my nausea and stomach, then my big concern was that I was only sleeping 4 hours per night, then my big concern was my anxiety levels and that my wife and I would get divorced because I was unable to go on vacations with her.

Then my thoughts were that I wasn't eating enough and my loss of appetite was getting bad. Then it was the depression with the crying fits for no reason that made me think. I freaked out even more when I was getting freaked out just being around my own family at Thanksgiving.

So looking back on the way I was, and seeing how I am now, I really owe myself a ton of credit. Having gone through all that stuff with no medication at all is impressive. I believe He knows that I've been through. One in 10 people have anxiety attacks which is 10 percent of the population. I'm not alone by any means.I'm sure even more people that quit smoking or chew have anxiety suddenly.

I'm getting better and better everyday at handling depression. For example, if I'm tired or depressed, I sat down or close my eyes for an hour and those feelings subside. If I've just had a big meal and my stomach feels sick, I go for a large walk to match it and feel better as a result. I have tools now!

I haven't been myself today but I was planning on this because of last nights nap. I only need to stay awake a little while longer. Grandpa said that taking naps isn't that great for you and says you are better off staying up until your regular bedtime. After the way I'm feeling now, I can't agree more.

Today I went to church and had a moment or wave of depression which lasted about 1 minute. It came with extreme tiredness. I let the wave ride out. We went outside and talked to the Priest a little bit after mass and then when we got home I had anxiety. I got anxiety because I was stressing out wondering if my wife would be upset with me if I wanted to just stay home all day. Heart rate — 116/83 pulse — 74 Symptoms today are tired, loss of appetite, slight depression, anxiety.

Today's anxiety range is a 5 out of a 10.

1/5/09, 9:12 p.m.

Today I attended the anxiety class. Even though I am getting better from taking the class, negative comments are still made at me the minute I get home about how I'll always have anxiety problems. I sometimes feel I'm up against my own family members with this. I don't care to have a whole lot of attitude today. This morning I had a hard time waking up and driving to work. I felt like a zombie. Learning to cope without nicotine in my life is really tough! I never thought I'd be going through all this. I just thought I'd quit, and be happy all of a sudden a week later.

At work I was fatigued off and on until around noon time. I took a nap at lunch. My level of anxiety today was around a 5–6 this morning and a 2 after lunch and a 1 after I finished the anxiety class. This morning I didn't have an appetite until around 11:30 a.m. I ate a little fruit and then lost the appetite until around 3:30 p.m. It wasn't until 8:30 p.m. that I felt I could REALLY eat something solid.

1/6/09

The day was better then yesterday's. I've been getting irritated with my wife today though. I started the day with a strange feeling in my head which I interpreted as depression. I was tired until around 11 a.m. Didn't have any depression again until 4 p.m. when I started thinking "bored" thoughts like "there's nothing to do at home" or "my life is so routine and don't have anything going on."

I have noticed that with each day that passes though I become less and less depressed. Or I understand how to handle it better. Boredom is part of my list of withdrawals now. My stomach has been good with indigestion stuff. Good thing is that the gas doesn't give me an upset stomach. My main issue now is the negative thinking that can instantly lead to depression and defeat. If I can get those thoughts under control some how, I'll be in good shape. I also want to start working out again and am planning for regular exercise soon.

1/7/09, 1 day less of 6 months off Chew

Tomorrow will make 6 months free from chew. Looking back on it I have no regrets. I'm glad I didn't go back and start chewing again. Today was an excellent day though. I was a little tired this morning until around 8 a.m. then I was fine after that. I didn't need to take a nap at lunch but I did anyways. I've been anxious to get better for a long time now. Next I'd like to start working on my self esteem again. Besides that, I noticed today that I started daydreaming about traveling again. It once again seems fun to me. If it seems fun to me everyday for the next 2 weeks, I'll take it seriously. I don't know if this is just because I'm having a good day today. Sometimes I get bummed out going to bed because I'm feeling normal by evening time most of my days now. I dread the morning because I know that with it will also bring fatigue, sadness and possible anxiety.

1/08/09, 6 Months Quit

Today made 6 months off chew. I felt good most of the day. I was in a positive mood until around 2 p.m. I had gotten back from lunch and was at my desk and became fatigued. I remained tired for the day but without the depression. We went to dinner at Chevy's and had some Nachos & hot wings. When the waiter brought my food I paniced for a second cause I thought "how can I finish all this food?" They don't allow take out for happy hour either! It lasted for a minute then the anxiety subsided. I feel God is watching over me. My prayers are being answered by Him. Anxiety level today was a 2 (scale of 1–10.) I am proud of myself for sticking through things.

I'm going to briefly summarize how the next 6 months went for me. Absolutely rancid! Everything was a task for me. I could barely enjoy doing anything except maybe going for walks, but besides that, I couldn't find fun in anything. Not even going to see my Grandparents! But like I stated earlier, this is what quitting is all about. If it was easy, then the chewing tobacco industry would be out of business because everyone knows this stuff is bad for you. We all know the harms or have an idea at least.

I had no idea that it would take this long or would be this severe though. I thought it would be gentler on me. I thought my body would pat me on the back for quitting and not resist so much. The key is, do not lose your mind during the 2nd have of your year quit. Stay focused, use your past journal entries as evidence that you have improved. Look back on those bad days you had and say to yourself, "Wow, I was pretty bad back then, I'm glad I'm past that stage in the quit."

7/08/09, (1 year quit)

Today makes my first year off chewing tobacco. I have learned so much about this nasty habit it's not even funny. I am committed to getting this book published so that young athletes don't make the same mistake I do. I always thought of myself as a easy going guy that's never bummed out for long periods of time. That is until I gave up chew. Mentally though, I can feel myself slowly creeping out of this logjam. Out of a 7 day week, I'll have 4 that are good, 2 that are okay, and 1 bad day. But my bad days really aren't as bad as before. I mean they still have the same sensations as before but I know how to cope with them now and combat.

Four or 5 months into the quit I had no ammo for this stuff and would just freak out and cry or feel disconnected from everyone. And the more attempts I made at trying to connect with life again, the more disappointed I'd become with myself.

An example of knowing that I'm improving is we went to dinner at a busy restaurant the other day. After eating I got over stuffed and began to have some sort of anxiety attack. But instead of freaking out and running for the hills, I sat it out and sipped my sprite. After a few minutes, it went away. I felt very good about this and was glad I didn't run to the bathroom.

I've learned that anxiety only has as much power as the attention that you give it. In other words, if you give it praise for what it's doing to your body and mind, then it's going to be very expressive... But if you pass it off as just a simple discomfort like it really is, then you are your way to recovery!

Chewing tobacco is what caused all of this garbage. I'm holding it 1000 percent responsible for what I've had to go through.

Learning to adjust to the new lifestyle

So if you took the time to read through my 6 month journal you'll find a wide range of emotions and things I had to put up with. I can tell you now, it wasn't easy. Let me remind you that I was never the type to get full blown depression, anxiety, stomach upsets or fatigue like this. In fact, I was a very healthy person pre-quit.

When a person quits chew it feels as though you've lost a person close to your heart. It's scary almost because you don't know when these emotions will pop in on you. One minute you can be happy as can be and the next you could be stressed out about something that makes absolutely no sense. Just like when my wife and I booked the trip to Belize and I had to cancel at the time because I didn't know what these emotions were all about yet.

I have learned thought that you can function under these conditions. Having the anxiety and depression is horrible yes and I wouldn't want my worst enemy to go through it but there is an end in site for those that decided to quit chew.

I can't say when it will go away either. This is just the first 6 months that I addressed here because at the end of 6 months were my "first signs" of feeling normal and being able to find joy in traveling, going out to dinner, or being with friends.

There will be times too when your friends and family may lose faith in you and possibly say negative things to you when they are disappointed in the new you. Avoid their non-sense. They don't know what's going on in your head and if they did experience these odd sensations they would probably be keeping their mouths shut or not giving their advice. I got a lot of the same comments like "Just try not thinking about stuff so much" or "You'll never be able to have a good time again, you are always depressed or having an issue" or "you just care about yourself and all your thoughts are about you, why don't you think about what other people want from you for a change"

Be prepared for more.I'm sure you'll have quite a list to share with others as well!

People tend to gripe when they don't understand something that others do. Even when I used to chew, people would say "why don't you just chew gum?" I would try to tell them that it's not the same and that this would satisfy me more then a piece of gum.

Understand that people just want the best for you. Don't get mad at them for giving you bad advice. Just say "thanks, hey I'll think about that." And then be on your way. If you are nasty with them, chances are, they won't be kind and thoughtful about your quit because of your attitude. You want a support system that is going to last you a year or 2.

There will be moments when you feel the world is totally against you and you feel like "giving up" on life and everyone close to you. They may be disappointed in how you are performing. It could be people from work that don't understand what's going on with you and expect you to be all happy 100 percent of the time or they could be someone from home that expected you to feel normal after a couple weeks.

I had one of these weeks not too long ago when it seemed like everyone close to me was disappointed in me about something. Each one expected a little more then usual. I felt like the center of the universe and everything revolved around me and my decision making. I kept looking into the future and saying to myself that there were even tougher days to come and this was a cakewalk. Anticipation is killer as Jeremy Kernion would put it. 99.9 percent of the time what you imagine in your head, never turns out to be that way in reality.

Try not to take this world so seriously if you can because you too will be checking out of this hotel someday too don't forget! The world will go on without you just fine.

Praying

I WAS NEVER ONE TO GO TO CHURCH OR PRAY a whole lot before quitting chew. But once I did quit and went to the Doctors a number of times with no answer as to why I was feeling the feeling I was, I had no choice but to pray.

If you aren't a religious person to begin your quit then there is a 99% chance you will be after all is said and done. I needed His help through all of this or I'd have ended up jumping off the Golden Gate Bridge. Just knowing that was available to me was better then any Doctor or Social Worker. I knew that if I prayed to Him every night to please help me get better and to stay with me during the withdrawal of nicotine, that I'd get a little bit better every day.

I didn't come from a religious family but my wife has a firm belief in God (in fact she's seen Him on the thrown and I believe her.) Just hearing that from her gives me the confidence to pray to Him. I know she doesn't lie about things either.

What happens during the first 6 months off chewing tobacco?

I can say that for a lot of people nothing at all happens. They for some reason don't go through any discomfort. Yes, I am jealous of them severely.

As you read in my journal, many things happened to me during my first six months off the chew. Probably the most difficult to overcome was the depression. Depression can really play tricks on the mind and cause other health problems. Depression isn't just feeling bummed out about something either. The way you can tell if you are having a depressive episode or not, is by trying to explain your feelings and symptoms during that moment. If you are unable to explain and feel its just way too complicated then you know that it is depression. I have a friend that is Bipolar that wakes up every morning to it and he doesn't start feeling good about his life until 5 p.m. everyday. I feel for him because he has to deal with this for the rest of his life.

He has a mental illness that he is being treated for in Bipolar disorder. You as a recovering chew addict are in a different situation then him. This is a temporary feeling of what my friend must go through for the rest of his life. If I knew when I quit chew that I'd face depression for as long as I did, I'd have never touched the stuff. But because I did it for so long (16 years) I couldn't avoid it. There is no way of preparing for it either. When the waves of depression come the one thing I can suggest is to lie down and close your eyes until the storm passes.

If you choose to stay awake and ride it out, do not take yourself and your thoughts seriously during these times.Your brain is still recovering and trying to make sense of all this without chew every day. Remember, how long did it take you to actually quit the drug? If you did it for many years, then plan for a lengthy withdrawal too.

Don't give yourself a timeline like I did either where you say, "well by 6 months off of chew I should be good and if its not then there must be something seriously wrong with me."

Quit because you want to

Do not quit chew because your Mom or Wife doesn't like that you do it. That would end up with you chewing twice as much and holding a grudge against them for putting you through all this. When you decide to quit, make it because you want to and not someone else.

When I decided to give it up it was because I panicked one day. I felt my heart racing uncontrollably and felt a constant sickness. It seemed to me like the chew was breaking down my immune system Helper-T cell per Helper-T cell. I was not a fan of this feeling. At the time I remembered saying to myself that it was all caused by the chewing tobacco. Why else would I feel that way? I was healthy, athletic, worked out on a daily basis. I noticed it was starting to give me heartburn too. Quitting may be the biggest decision of a person's life because of how much control it has over your mind. When you decide to toss your can for good, feel good about it. Sometimes the Kodiak bear needs to frighten us a little (ever wonder why it is they put a bear on the cans?) It had to scare the daylights out of me before I was able to quit.

Wonder when the drama will end?

Get ready for the rollercoaster ride of your life the minute you put the chew can down. As I said earlier, you may be one of those lucky ones that does not have a problems when you quit and you feel great. Congrats to you. But if you are in the other 99% that has at least 1 withdrawal symptom then be prepared to have moments of extreme emotion as I did in the journal that you read. I can say that a good thing to do is to become a member on the about.com forum for smoking cessation. Even though you may not have been a smoker, they too are quitting nicotine so they will be able to relate to your feelings and frustrations along the way. They have a deal too where you get pats on the back for the milestones you reach like 1 week quit, 1 month quit, 6 months, 1 year. Some people go on there that have been done with nicotine for several years and they take the time to share their struggles with others on there that are in the early stages of their quit.

For me, it was always comforting to read others stories they had and how much they too struggled.

The Big Breakup

I LIKE TO THINK OF QUITTING CHEW like your first real breakup with someone. When you first fell in love you may have thought you'd be together forever and that nothing could come between the two of you. But as time went on in your relationship you found that some things you didn't like about that person that bugged you. You would think about how much you wanted that person to change but you knew that they wouldn't change their ways, it's just the way it would have to be if you were to continue the relationship. The longer you stay in the relationship with the wrong person, the longer you suffer and don't live a quality of life. In order to break things off with them, it's going to take courage, strength, knowledge, and a knowing of how good the end result will be.

When you break up though, initially it can be tough and you feel your emotions can get the best of you the majority of the time. You wonder if you'll ever be a normal person now and that you have made a mistake by breaking up.

This is all the same idea when a person breaks up with chewing tobacco. They, at first love everything about it, it tastes good to them, satisfies, keeps the person interested in life. Then as time goes on they notice that "every once in a while" my stomach will feel funny or "my teeth ache in the morning." Suddenly you are finding out more and more about the Kodiak Can then you thought. You begin asking questions to others that have had a relationship with the Can at one time or another and they had health issues from it too.

Finally the day comes that you can no longer stand the tortures that you are going through with chew so you decide to stop. Think of it just like that break up I talked about and be prepared for a "break up style" recovery. Like a Southwest plane in the back seats during take off, there will be turbulence!

Fake Mint Chew

There are a few companies that sell "fake chew" that is made from mint leaves not tobacco. I have tried the stuff myself and actually would mix the real stuff with the fake just before I officially quit cold turkey. Although I don't advise using fake chew as you still continue a habit of opening a can and pinching and placing something in your lip, it's still considered a healthy alternative to chewing tobacco. I used to contact a place in Oregon called "The Oregon Mint Snuff Company." The number for them is 1-800-EAT-MINT.

The other mint chew company is called Hooch Snuff. I haven't ever tried them so I can't say what they are like. My experience to the mint snuff was good, though. You won't get the same satisfaction as real chew but it may eliminate half the addiction of the act of doing it.

Avoid the Partyman

Yes stay away from parties or your friend the Partyman. He the one that knows where all the great parties are and he wants to party hard with you. He will talk to you about all the good times you both had together at Lake Wombusok or where there will be a great party coming up… Why? Because he's the Partyman and he tells you that you're top priority!

Avoid this guy and Parties for a while, a long while. What will happen is you'll go there and be a little nervous maybe because you won't have your chew that you normally would. So instead you'll have a few beers or screwdrivers. Then over the course of the night you will feel more laid back and say "hey maybe a chew wouldn't be that bad right now, its only 1 anyways right?" You'll end up heading down to the liquor store with a few others that are probably going to get cigarettes.

The next day you will feel rotten, I can guarantee it. Why? The reason why is because you lost your quit which may bring a little depression because of how you let yourself down after all that hard work. And also because of how powerful this chew was. Your body wasn't used to it just like when you first chewed so it could have given you a stomach ache for the night causing you to stay restless.

Point is, there will always be parties, and the Partyman can still go to dinner, play basketball, or play pool with you. You just have to tell him that you need his help in not offering any parties for now. Let him know you are trying to get off the chew habit because you don't want to get cancer and in order to do this, you have to avoid certain things. But tell him, you'd like to work on your health and would like to maybe start lifting weights with him if he's interested in that sort of thing. Who knows, maybe you can help the Partyman out too and get him on a health kick.

Traveling

I know that this was a very hard topic for me. Traveling originally was something that I enjoyed and when I saw a plane in the sky I'd always wonder what vacation destiny it was headed for. I would fantasize with a chew in my mouth about how great it would be to head over to a Sandals Resort and snorkel and fine dine with my wife.

When the time came to stop chewing, that all changed. For some reason or another, my mind said no to traveling, no to getting away, and no to any fantasizing about trips. When my wife would even suggest going out of town for the weekend my body would go into a strange state where I couldn't relax because I was so caught up in how dreadful my trip would be. I'd speculate how we'd travel to a far away country and I'd find myself sick to my stomach, depressed and having anxiety attacks with no where to run. I just could not relax. I felt like a chronic worrier. And if I backed out of wanting to go on a trip with her then I'd get sick to my stomach anyways at home because I was then worried she was all upset with me for being "boring."

I'd say the only solution to this state is to ride it out. I don't suggest going far away from home since you probably won't have a good time (when you are between 1–6 months into your quit.)

Little trips are good though to those weekend getaway's. For whatever reason, I feel quitting chew plays a roll in the way you feel about certain things. Once you are past a year or so you will enjoy these resorts again so don't worry, you are just going through something right now. If you are reading this book and you are 1 day—5 months and you are housebound, attempt a weekend trip where you are the driver, that way you are in control. If you are 5 to 8 months in your quit, try another weekend trip only go 100 miles farther. Then when you are 8 to 9 months in, do a plane trip to the closest airport from yours and book a hotel room. It's called "desensitizing" and this will get you back to doing the trips you enjoy.

You will enjoy trips without substance and impurities in your body and you'll be able to do all the things you used to do when you were a chewer only you'll having even more fun cause you'll have more money and time. You will have more money if you start a money market account and put $4 per day like I do. This is $4 that would have been spent on chew otherwise.

You'll have more time because you used to always have to take "timeout" to have your chew away from people in your private place. You may not realize it but if you had chewed 4 dips per day at 30 minute episodes for the week of your vacation, that's 14 hours or nearly 2 days of your trip! What a waste if you ask me.. It seems like such a big priority with everything you do when you are addicted. It would go good with just about anything you can think of. "Oh it will be nice to sit out on the beach and have a chew" or "when we do our nice drive today, we can enjoy the scenery and have a chew" it's totally unnecessary.

I was reading an article not too long ago that said that chewing tobacco is more addicting and accepting to the brain then is heroin even. That is nuts but very true from this study they did.

Learning to live without chew

As you read in the 6 month journal of mine, it can be a major pain to stop chew. Not so much the "cravings" of wanting to have a chew but the mental problems that come after you quit for a long time. If you can somehow remember, try to tell yourself that these are all good signs that you are having and that its part of the healing process, you will speed up your process. I doubted myself halfway through and just couldn't grasp that all these thoughts and sensations were from me getting off the chew.

I was under the impression that quitting chew would only take a couple weeks and then I'd feel normal. I'd be able to sleep good, smile, be happy all those good things but no….I had to go through the dark creepy tunnel alone for many days hoping to catch a glimpse of a light at the end coming. As Winston Churchill once said, "If you are going through hell, keep going." And that's exactly what I intended to do.

Besides telling my mind, it helped going on the about.com quit smoking forum. Reading about others and how screwed up they were and how deep they were in their own quits made me realize that there were others feeling these feelings too!

The other thing is you will get a lot of, "I think you should see a doctor because you aren't acting right" and a lot of "Maybe you should see a specialist" from people that don't know what they are talking about. These types of comments can either screw you up bad or do nothing to you. When I first quit chew I didn't know what to expect so everytime I'd tell someone about my symptoms, they'd get me all freaked out and say you need to see a doctor right now!

I had no idea that the symptoms I was having were part of the recovery process. So after a million doctor's visits to about 6 different people I finally decided I wasn't going to tell non-professionals about my problems anymore. I was going to keep it bottled up inside me for people that absolutely knew what they were talking about.

Foods to eat and avoid

Certain foods I learned I could no longer consume once I quit using chewing tobacco. Before I quit chew I could eat anything in the world and not have a problem. But for some unknown scientific reason I couldn't consume certain foods. The real foods that caused me to have severe indigestion were Garlic & Chocolate. Either of those 2 would have me on the toilet before long. In the beginning stages of my quit I also had a hard time consuming anything with lactose. I had made sure that I wasn't allergic to anything by getting an allergy test. The test came back saying that I had no known allergies to anything.

For your own benefit, avoid the following foods for the first year of your quit:

Chocolate
Garlic
Milk & dairy products
Onions
Bell peppers or hot peppers
Deep fried foods
Bell Peppers
Peas, Greenbeans
Salsa

All of these cause major stomach upsets after quitting chew at first. I know this from my own tests. Later in life you'll be able to consume these foods normal again but for now, eat good clean foods. Keep getting your fiber daily too. I found that consuming a lot of fish and chicken was great for me. But I avoided eating salmon after 6 p.m. because it's so dense in protein that it caused me to have nightmares at night.

I modeled for 10 years so I always had to eat healthy and stay in decent shape so it wasn't difficult for me to follow these guidelines. But if you are one that likes to binge eat, your stomach won't tolerate it. It's adjusting to the new you and life without chew also so you've got to eat small portions of food through out the day. Foods I do recommend are:

Veggies
Granola
Soups such as chicken noodle
Chicken (peel the skin)
Fish (omega 3 fatty acids/ good for depression)
Plenty of fluids
Spring Water
Shell Fish

Replace coffee and caffeine Soda's with water. You can live without that stuff I'm sure. You'll feel good consuming just water all the time for your beverage too. One thing to avoid is drinking water or fluids after 7 p.m. Maybe you're like me, you'll end up having to get up and go to the bathroom a few times a night if you drank too much water.

I've also learned that if you buy just the purified water at the grocery store, it's okay to drink but has nothing good for you in it. If you buy Spring Water like Crystal Geyser, Alhambra, Arrowhead, and a few others you are getting water that has been oxygenated and full of minerals. That's the water you want to be consuming into your body each day from here on out. Tap water is no good and for me, it would cause upset stomachs.

Chew is becoming Popular

Chewing tobacco is rapidly becoming a huge market. Polls show that it is over 3 times as popular as it was 10 years ago. At almost any gas station or liquor store you can find different brands all behind the store clerk.Chewing tobacco is getting more expensive by the minute and is now around $5.50—$9.80. I've even driven to some remote areas where they were selling cans of Copenhagen for $10 a piece! A $10 a day habit can get extremely expensive.

For my honeymoon we stayed at the Sandals Resort in the Bahamas and I ran out of chew. I went down to the gift shop and they were selling cans of Copenhagen for $15 American dollars. $15 for a can that will last 24 hours!

Here are some figures I want to share with you on how well Phillip Morris and Reynolds Corp. are doing from Smokeless tobacco sales. If I had a more current graph or pie chart to show you, I would but this is the most modern I can come up with since nearly all studies are on smoking and not chew. You get an idea of where we are headed with this table below. You can just imagine how many more people chew now a day's as compared to 1995:

SMOKELESS TOBACCO SALES BY YEAR
(In Pounds)

Year	Total
1985	121,449,115
1986	118,778,334
1987	116,540,281
1988	114,433,782
1989	116,440,365
1990	117,415,326
1991	120,110,686
1992	118,372,693
1993	115,888,785
1994	115,495,201
1995	116,387,464

SMOKELESS TOBACCO SALES BY YEAR
(In Dollars)

Year	Total $
1985	730,618,970
1986	797,777,885
1987	852,717,347
1988	901,654,382
1989	981,637,304
1990	1,091,170,201
1991	1,237,961,670
1992	1,361,360,729
1993	1,475,460,518
1994	1,612,098,989
1995	1,735,840,489

SMOKELESS TOBACCO ADVERTISING AND PROMOTIONAL EXPENDITURES BY CATEGORY

	1985
Television and Radio Advertising	$26,584,731
Total Print	$ 8,719,379
All Other	$44,764,119
Total	$80,068,229

	1986	1987
Newspapers	$626,979	$1,452,710
Magazines	$6,226,654	$9,237,988
Outdoor	$2,722,557	$8,321,315
Television and Radio[(16)]	$16,067,211	$0
Audio, Visual	$2,579,268	$0
Transit	$0	$0
Point of Sale	$4,234,207	$5,789,436
Promotional Allowances	$8,231,580	$7,554,592
Sampling	$13,699,156	$13,877,923
Distribution:		
Bearing Names	$2,353,816	$4,312,094
Direct Mail	$20,844	$48,979
Public Entertainment	$13,823,266	$14,844,425
Endorsements	$435,710	$245,105
All Other	$5,655,458	$2,092,477
TOTAL	$76,676,706	$67,777,044

SMOKELESS TOBACCO SALES TOTAL ADVERTISING EXPENDITURES (In Dollars)
TOTAL ADVERTISING

YEAR	REVENUES	EXPENDITURES
1985	$730,618,970	$80,068,229*
1986	$797,777,885	$76,676,706
1987	$852,717,347	$67,777,044
1988	$901,654,382*	$68,223,671
1989	$981,657,304*	$81,200,611
1990	$1,091,170,201*	$90,101,327*
1991	$1,237,961,670	$104,004,040*
1992	$1,361,360,729	$115,346,708*
1993	$1,475,460,518	$119,230,826*
1994	$1,612,098,989	$125,972,408
1995	$1,735,840,489	$127,323,282

*Information Courtesy of Phillip Morris USA

Looking Forward

NOW THAT YOU'VE LOOKED AT THE TABLES and seen what they put into advertising and marketing, you can see what their bottom line is and how much money the tobacco makers really make. If you notice there aren't any chewing tobacco commercials on American TV stations, it's because many years ago our congress passed a bill that doesn't allow that to happen. It's also why in the beer commercials you don't see anyone actually taking a sip of the drink. They hold it up to their lips, or salute or they show the beer flowing over the glass. That was part of the bill.

The way they advertise though is through magazines, particularly sports and outdoors magazines. This is where they are allowed to advertise and they do so, heavily.

What's good though is they are forced to have a part in the advertisement that warns you of the product and explains how it can cause tooth loss or cancer.

Closing Words about Chew

I was reading a book on nicotine and the harms they cause on our minds and bodies and learned from it that chewing tobacco not only causes gum disease but also can cause Lung

Cancer, Stomach Cancer and even Liver cancer. There is a lot I have learned from writing this book. Chew is a poison on the mind and body. It holds you back from all the joys in the world and family, family that won't be around forever here on earth. You'll miss those moments one day when you passed on hanging out with your Grandpa because you decided to have a chew that day instead.

Remember this stuff is a lot harder to quit then regular people quitting smoking. Most of them averaged a couple packs a day for their quits. As I was saying earlier in the book, the nicotine content in a can of dip or chew is approximately 144 milligrams, which is equal to about 80 cigarettes. In other words, one can of snuff equals about 4 packs of cigarettes a day. That meant for me I was this way for 16 years.

I don't know too many smokers that are able to find time to smoke 4 packs a day. That would be nuts. So remember to be patient with yourself because you're getting used to not having large doses of nicotine being put in your blood on a regular basis anymore. This will take time, a lot of time.

I've also learned that there is not one thing good about chew, not one! Even the feeling or high I got from it when I initially did it would always turn into a sluggish, lethargic sensation so not even that was good.

I've learned that over the long run, it slowly eats at the mind like cancer. You think a little slower, react slower, you have less tolerance for people, less patience.

Looking forward now that I've quit chew, I will never go back to this stuff. To me, it's just like waking to a nightmare everyday. If it seems good to you now and you've just started a year ago please take my message seriously. I can't count how many trying days I've had since chew had become a part of my life. Take care to all of you and yes, you can quit this stuff! Troy Aikman even quit.

Links for chewers:
Quitsmokeless.org
Whyquit.com
Nstep.org
About.com (nicotine withdrawal)

In the long run the benefits of quitting chew will be great but in the first 1 and a half years it may be dreadful. As I write this its been nearly 1 yr off chew and still to this day I'll have some very strong symptom moments that I don't see how things will ever be better.

Remember, you are readjusting to this new life without tobacco. It had a grip on you for many years and was there for you whenever you needed a friend. When you were tired, it was there, when you were awake and wanting someone to be with, it was there, when you were down, it was there, and when you'd celebrate, it was there. It's like a best friend of 16 yrs for me that I basically told to get lost.

So even though I say out loud that I can't stand chew and will never go back to it, my body and mind are still sorting out the grieving process.

Out of all the symptoms I've had, the hardest has been the depression followed by the fatigue and anxiety after that. Remember people have anxiety when they think about the future and have depression when they think about the past. That's how it works.

It's hard to get anxiety over something that already came and went right? I can't remember anything that I've stressed out about like that. Same with depression, I don't get depressed over something that hasn't arrived yet.

Your 1 year anniversary

Congrats on making it one year! You have reached your key as they'd say on about.com's smoking cessation discussion group. There are not a whole lot of you that make it to this point so you should feel proud of yourself. A lot of people start off strong on their quits only to get too confident in thinking they can just have a drink or a smoke or a chew after a couple months. They eventually fall victim to whatever their addiction was.

I read peoples messages on certain boards saying how they've been quit from smoking for 90 days but in the last week they've "allowed themselves" to have light cigarettes and at least those aren't what they were smoking.

Sure I give them a little credit for going the 90 days and also for not just going back to the hardcore smoking habit they once had, but what did they accomplish by doing lights? It's still a cigarette. Those people aren't going to get as much enthusiasm from me as those that said enough to smoking once and for all. They didn't look back, they didn't miss it.

It's just like the guy that tells you that he was addicted to multi vitamins and would take one every single day even though the Doctor said not to. So he stops but then would go and start taking Children's chewable vitamins instead.

Yes, they are less potent with a lower dosage of supplement in them. But a vitamin is a vitamin is a vitamin!

I'd go with a sense of responsibility to quit this habit. Nicotine is extremely potent and addicting and ranks right up there with alcohol.

My friend Rich has been quit from chew for 4 years. He went through some of the same drama that I went through in quitting nicotine. We both did about the same amount per day for the same amount of years. I recently contacted Rich to ask him some questions about how he felt about quitting chew and his reactions:

1) When you first quit chewing tobacco, say your first month, describe what you went through as far as symptoms?

I was extremely irritable, nervous, and anxious. I suffered from sporadic headaches, stomach pains, and irregular bowel movements.

2) So during your quit, how were you treated by your family, friends? Did anyone feel you suddenly had mental problems?

People around me were generally unsympathetic and thought my reactions or symptoms to nicotine withdrawal were psycho simatic, not real, or exaggerated. As the days passed and my feelings or symptoms didn't go away I could sense that people close to me may have thought the problems were mental.

3) When you chewed tobacco how was your lifestyle different then it is today being off it for 5 years?

When I chewed tobacco I started at a young age being a teenager. I was single and lived life carefree. Today, being off chew now for 5 years my life is drastically different being married with three young children. I now have a different perspective on life, it is not just me now, I have my wife and children that depend on my health and wellbeing. When you become married and have children I found myself taking less risk, and also taking other peoples feelings into consideration.

4) Was there anything positive that came out of chewing tobacco?

Yes. I would say that chewing with close friends developed comradery and strengthened relationships and bonds. The friendship and bonds formed over in depth conversations whether in small groups or socially made for memorable moments that I fondly reflect on years later.

5) When you did decide you wanted to quit chew, what were the deciding factors?

First and foremost, my health and wellbeing. My fear of Cancer and feeling that chewing daily for 15 years made it inevitable that I was going to eventually get sick. The day I decided to quit chew was a day I will always remember. It was in the morning and I brushed my teeth and something told me to check my gums in the mirror. When I lowered my bottom lip I remember seeing a large white spot that couldn't be wiped away. At that moment in time I immediately thought the worst, my fears of Cancer became real. It turned out that it was pre-Cancer and I was O.K. But from that day forward I vowed never to chew again. My fear was that if I just continued to chew and didn't use this incident as a learning lesson, next time I may not get a second chance.

6) What long term health problems have you discovered now that you've been off tobacco?

My lower jaw gums are completely receded, to the point that the tooth root or nerves are partially exposed. This condition makes my gums extremely sensitive to food or beverages. I may eventually need gum surgery to repair the damage caused from past chewing. The other long-term issue is mentally, as far as brain chemical imbalances that now have to be regulated by prescription medications. You see when I chewed it was not only addictive but it calmed me down and was a sense of relief or calmness to me. It was an outlet I used to unwind and relax, the same way that people have a beer or glass of wine after a long day at work. My mother smoked will she was pregnant with me, when I was born and grew up I was subjected to second hand smoke from both of my parents, then I started chewing when I was 15 years old and didn't stop until I was 30 years old. That equates to 30 years of nicotine exposure. When I stopped chewing my body and mind totally went into shock and I began to experience severe anxiety and occasional panic attacks for no reason that required visits to the ER. All of the sudden out of the blue I thought I was dieing or sick from some rare disease, it was totally psycho simatic, but in the moment it was very real to me. I would get so worked up for no reason my heart would pound and I thought a heart attack was immanent. I didn't have chew to calm me down, so I would get so worked up that panic attacks set in. I had to take anti-anxiety medicine to regulate my brain and thought process until I realized that I am O.K. I don't get worked up anymore or feel anxious because of medicine now, not chew.

7) Do you ever crave chewing tobacco and when was your last craving?

I don't crave chewing tobacco at all. My last craving was over five years ago. I don't crave, but really miss having a chew or the oral fixation of chewing during long road trips or when I am working alone or watching TV at night. I don't miss the nicotine, just the process if that makes sense.

So there you have it, I hope that this book will help guide you in the right direction with successfully quitting chewing tobacco. It can be done, you just have to be patient with yourself and not get frustrated.

Of course that's easier said then done as there will be times of extreme frustration. Make sure to keep moving, keep active. Plan on feeling rotten for about a year if you are an extreme chewer like I was that did a can per day.

This goal can be beat. You can beat the bear! But it has to be because you want it and not because someone else does.

When you are ready, contact your doctor and let him/her know you'd like to get off chew. See if they have any advice to give you or if they are completely clueless. Most know their stuff. I was lucky in getting Dr.Ryan A.Gorman who came out of Boston College. He knew this chemical nicotine and knew what I was dealing with and actually coached me through my quit via email. I can remember emailing him way out at 8 months quit asking if my symptoms were all from quitting nicotine and he quickly responded with a "yes" confirming my suspicions that nicotine is a very powerful drug that can not be taken lightly.

When you go to your Doctor make sure to tell him/her how much you chewed and how you'd like to quit. They may recommend the patch or Welburton.

While I didn't touch any crazy pills, I don't totally count them out. They can help you.

Probably is they are habit forming so you could be getting off chew just to get hooked on a pill instead. Therefore, if you are serious and know that you are in for a battle ahead, you can do this cold turkey.

Your body can take this punishment which is a lot healthier for you then being on snuff.

Things are looking better everyday! The stock market is even on its way back up! Today is 10/15/09 and the Dow is at 10002.13. I only see things from a positive light now which is so much easier to deal with in life.

LaVergne, TN USA
17 November 2009
164326LV00004B/6/P

9 781602 644786